MEDIATIZATION AND MOBILE LIVES

Mediatization and Mobile Lives: A Critical Approach contributes to a complex, situated and critical understanding of what mediatization means and how it works in contemporary life.

The book explores the tension between the extended capabilities offered by media technology and growing media reliance, focusing particularly on mobile middle-class lives. It problematizes how mediatization is culturally legitimized in our times, when connectivity and mobility are increasingly seen as mandatory elements of self-realization.

Supported by extensive fieldwork carried out in contexts of gentrification, elite cosmopolitanism and post-tourism, André Jansson advances a critical, cultural materialist perspective of mediatization as he examines how people are torn between the new opportunities afforded by their mobile lives and the feeling of being trapped by our connected media culture.

Mediatization and Mobile Lives offers an engaging and critical exploration of the interplay between mediatization, individualization and globalization, making it an ideal resource for students and scholars of Media and Communication.

André Jansson is Professor of Media and Communication Studies and Director of the Geomedia Research Group at Karlstad University, Sweden. His most recent publications include *Cosmopolitanism and the Media* (2015, with M. Christensen) and *Communications/Media/Geographies* (2017, with P. C Adams, J. Cupples, K. Glynn and S. Moores).

MEDIATIZATION AND MOBILE LIVES

A Critical Approach

André Jansson

Routledge
Taylor & Francis Group

LONDON AND NEW YORK

First published 2018
by Routledge
2 Park Square, Milton Park, Abingdon, Oxon OX14 4RN

and by Routledge
711 Third Avenue, New York, NY 10017

Routledge is an imprint of the Taylor & Francis Group, an informa business

© 2018 André Jansson

British Library Cataloguing-in-Publication Data
A catalogue record for this book is available from the British Library

Library of Congress Cataloging-in-Publication Data
A catalog record for this book has been requested

ISBN: 978-1-138-72362-7 (hbk)
ISBN: 978-1-138-72363-4 (pbk)
ISBN: 978-1-315-19287-1 (ebk)

Typeset in Bembo
by Deanta Global Publishing Services, Chennai, India

CONTENTS

ILLUSTRATIONS

FIGURES

TABLE

ACKNOWLEDGEMENTS

Mediatization and Mobile Lives is an attempt to bring together research projects and theoretical ideas I have worked on for more than a decade. As such, I hope that this book will succeed in presenting a relatively broad picture of what it means to live in a mediatized society. I also hope that the analyses will provide an inside view of how mediatization works and is experienced in mobile middle-class settings. The empirical material spans several contexts and includes sixty interviews in total; these are now brought together for the first time within a joint theoretical framework.

I am very grateful for the project grants that have made it possible for me to carry out this multi-sited research and explore various aspects of mediatization over the years. The projects that I have participated in, and in some cases led, are: *The Post-Industrial City: Culture, Identity and Life Forms* (funded by the Swedish Research Council, 2003–5); *Rural Networking/Networking the Rural: Participatory Culture and Civic Communities in the Swedish Countryside* (funded by the Research Council Formas, 2008–12); *Secure Spaces: Media, Consumption and Social Surveillance* (funded by the Swedish Research Council for the Humanities and Social Sciences, 2009–12); *Kinetic Élites: The Mediatization of Social Belonging and Close Relationships among Mobile Class Fractions* (funded by the Swedish Research Council, 2012–15), and *Cosmopolitanism from the Margins: Mediations of Expressivity, Social Space and Cultural Citizenship* (funded by the Swedish Research Council, 2012–16). I want to thank all the interviewees who kindly participated in these projects, and a special thanks to Sandra Engelbrecht who helped me with contacts during my fieldwork in Geneva in 2014. I also want to thank the good colleagues I have had the privilege to work with on these projects: Magnus Andersson, Miyase Christensen, Karin Fast, Jenny Jansdotter, Thomas Johansson, Johan Lindell, Linda Ryan Bengtsson, Ove Sernhede and Tindra Thor.

There are also other academic platforms that have been important for discussing and thinking through the arguments of this book. From 2012 to 2016 I was a member of the Sector Committee on the Mediatization of Culture and Everyday

Life (supported by the Swedish Research Council for the Humanities and Social Sciences), which organized several successful workshops and colloquia on mediatization. I am grateful that I was given the opportunity to be part of such an intellectual milieu, and want to thank the other members of the committee for many rewarding discussions: Göran Blomqvist, Mats Ekström, Johan Fornäs, Anne Jerslev, Ulrika Knutson, Pelle Snickars, Eva Swartz Grimaldi and Maria Wikse. Since 2013 I also been the director of the Geomedia Research Group (partly funded by the Swedish Research Council for the Humanities and Social Sciences) at Karlstad University, which constitutes my everyday work environment and a vital source of inspiration. I want to thank *all Geomedians* for producing such a hospitable space for the exchange of ideas.

The synthesizing nature of this book encompasses many ideas, as well as the underlying empirical material, that have been presented and discussed at conferences, workshops and meetings over the years. I have received valuable input from a great number of people. I am particularly grateful to those who have given explicit feedback on various parts of this book or invited me to give talks or participate in other academic exchanges related to this area of study: Paul C. Adams, Stina Bengtsson, Felix Bühlmann, Julie Cupples, Dana Diminescu, Kevin Glynn, Annette Hill, Stig Hjarvard, Bengt Johansson, Maja Klausen, Knut Lundby, Peter Lunt, Ulf Mellström, Shaun Moores, Maria Månsson, Kristian Møller Jørgensen, Zizi Papacharissi, Marcus Prest, Toke Riis Ebbesen, Scott Rodgers, John Tomlinson and Mekonnen Tesfahuney. To some extent this book also comprises discussions that have appeared in previous publications. These are:

Jansson, A. (2014). Indispensible things: On mediatization, space and materiality. In Lundby, K. (Ed.) *Mediatization of Communication (Handbook of Communication Sciences, Vol. 21)*. Berlin: De Gruyter Mouton. (Parts of this chapter have been reworked and incorporated in Chapter 3.)

Jansson, A. (2015a). Using Bourdieu in critical mediatization research: Communicational doxa and osmotic pressures in the field of UN organizations, *MedieKultur* 58: 13–29. (Parts of this article have been reworked and incorporated in Chapters 2 and 5.)

Jansson, A. (2015b). Interveillance: A new culture of recognition and mediatization, *Media and Communication* 3(3): 81–90. (Parts of this article have been reworked and incorporated in Chapter 4.)

Jansson, A. (2016). How to become an "elite cosmopolitan": The mediatized trajectories of UN expatriates, *European Journal of Cultural Studies* 19(5): 465–80. (Parts of this article have been reworked and incorporated in Chapter 5.)

Jansson, A. (2017). Critical communication geography: Space, recognition and the dialectic of mediatization. In Adams, P. C.; Cupples, J.; Glynn, K.; Jansson, A., and Moores, S. (Eds.) *Communications/Media/Geographies*. London: Routledge. (Parts of this chapter have been reworked and incorporated in Chapter 4.)

Finally, I dedicate this book to my parents, who have always supported me in my mobile endeavours, near and far.

André Jansson
Kristinehamn
31 January 2017

1

INTRODUCING CRITICAL MEDIATIZATION RESEARCH

In June 2014 I interviewed a Finnish man working for the United Nations (UN) in Geneva, Switzerland. I wanted to know more about how the rapid expansion of digital media had affected the working conditions and life environments of people with highly mobile and international careers. At the time of the interview, this man – we can call him Ruben – was in his early sixties and had held various positions within the UN system as well as in other internationally oriented organizations. From the late 1980s his career was marked by intense travelling, which was required for the kind of work he was interested in. During the 1990s he travelled about one hundred days every year and visited sixty countries. But things changed. Today he holds a specialist position and tries to stay put in Geneva as much as possible.

> When I started travelling in 1989 it was much more pleasurable. In 1989 we made up a programme via letter-writing or using telex, and then there were always a few meetings that ended up not taking place, and in the evenings I was free, didn't have any mobile, no laptop. If I was away for two weeks I phoned the office perhaps once a week to ask if everything was ok. But now, you are expected to do the same work while travelling as you would have done if still in the office. That's a bit strange…

Let us reflect on these words. Ruben says that the everyday saturation of media and communication technologies has made professional travelling less "pleasurable". What does this mean? We could of course argue that international travelling in the past was a more luxurious activity reserved for individuals occupying status positions, and that such travelling included services as well as "free time" that made life comfortable and interesting beyond the meeting schedules. In times of mediated connectivity, travelling can no longer be an escape. We could also argue that media are now helping professional travellers stay in touch while on the move, enabling them to work and communicate more efficiently and thus having less to catch up on when returning to the physical office. Still, Ruben's story invites us to consider the possibility that the continuous access to digital workspaces and communication channels leads to information pressures and social expectations that are difficult to cope with. Such developments may involve existential and social costs when individuals feel that their resources are exploited and other areas of life are fragmented or pushed aside. Ruben has in fact given up some of his earlier career ambitions. And he complains that his employer still does not provide him with a smartphone even though he is often expected to be reachable beyond stipulated office hours.

This is just one example of the ambiguous nature of mediatization. I have chosen Ruben's story because it points both to the general features of mediatization and to the particular approach that I want to advance in this book. On the general level, Ruben's experiences illustrate what many theorists today have come to describe as *mediatization:* a historical meta-process of structural transformation pertaining to a variety of social and cultural realms, conditioned by altered forms of mediation. In the introductory chapter to the edited handbook on the *Mediatization of Communication* Lundby (2014a) concludes that whereas mediation refers to "regular" forms of communication involving some kind of vehicle or medium, mediatization points to the broader "transformative" consequences of such processes. Mediatization is thus to be understood as a structural concept, referring to overarching societal transformations that in themselves contain altered and socially shaped forms of mediation. Along these lines Couldry and Hepp (2013: 197) state that movements of mediatization "reflect how the overall consequences of multiple processes of mediation have changed with the emergence of different kinds of media". Similarly, Krotz (2014: 137) holds that mediatization is a meta-process comparable with other meta-processes (such as individualization, globalization and commercialization); that is, "a long term development that includes media change and the respective change in culture and society". Hjarvard (2013: 1), in turn, begins his book on *The Mediatization of Culture and Society* by stating that "the concept of mediatization has proved useful to the understanding of how the media spread to, become intertwined with, and influence other fields or social institutions".

Behind these formulations (and other attempts to define mediatization) there are still disagreements and controversies regarding the more precise epistemological status of this meta-process. There are, for instance, tensions between "institutionalist" and "social constructivist" strands of mediatization researchers (see Couldry and Hepp, 2013), as well as between those who posit mediatization as a "paradigmatic turn" within media and communication studies (Hepp et al. 2015; see also Lunt and Livingstone, 2016) and those who call for an "open agenda" (Ekström et al., 2016). There are also those who question the validity of the concept altogether (e.g., Deacon and Stanyer, 2014, 2015). Still, the general orientation and justification of the term seems to gravitate around the fact that modernity encompasses social and cultural changes in which more and more areas and forms of practice become saturated with and adapted to media technologies and institutions. The gradual normalization of mediated communication affects leisure time as well as work time, public places as well as private places. It affects politicians, tourists and single parents as well as mobile professionals within the UN system.

So what is it I want to achieve more specifically in this book? Let me state initially that mediatization research, in my view, neither can nor should be envisioned as a "new paradigm". I believe that mediatization, especially given the complexity of this type of phenomenon, should be analysed from a variety of perspectives and paradigmatic viewpoints (see Ekström et al., 2016). Having said this, I find mediatization a very useful term for addressing and specifying media's broader significance in culture and society.[1] My aim is to advance a complex, situated and critical understanding of what mediatization means and how it works under modern life-conditions. While mediatization as such refers to a complex meta-process, on a par with, for instance, individualization and commercialization, it can also be broken down into sub-processes, which in their turn can be operationalized and studied empirically in particular time-space settings in order to illuminate mediatization's multiple and contextually formed expressions (Krotz, 2014: 148–53). The case of Ruben's altered experience of work-related mobility is one example.

But Ruben's story also illuminates what I see as four interconnected weaknesses of contemporary mediatization research. The first weakness springs from the above-mentioned nature of mediatization as a complex and inherently contradictory meta-process; that is, there is still a lack of *critical perspectives* on the ambiguous consequences of mediatization, especially in relation to everyday life. Ruben's life biography has evolved in close relation to various media developments in which the uses of technologies have, on the one hand, enabled certain forms of communication at-a-distance, and thus the kind of mobile career that he once strived for, while on the other hand invoking experiences of stress, restraint and intrusion. At a certain point the pressure of mediatization even reached a tipping point so that travelling was no longer associated with pleasure and was thus avoided rather than desired. A key argument of this book (developed below and in Part I) is that mediatization entails *a dialectical relationship* between liberating forces and increasing socio-technological dependence. This ontological understanding of the basic nature of mediatization, in turn, corresponds to an epistemological approach built around *immanent critique*. Mediatization research should try, in a more elaborated way than has so far been the case, to unveil the inner tensions, ambiguities and contradictions of a society in which media technologies have become taken-for-granted parts of everyday life (to different degrees in different groups) and in which social autonomy is at risk. This is not to say that mediatization is something inherently negative; just that the concept may assist the social sciences in framing, naming and specifying the social discrepancies that evolve over time in relation to media change.

The second shortcoming concerns the *lack of specificity* in much mediatization research when it comes to demarcating what are (or are not) to be understood as elements and articulations of mediatization (see also Ekström et al., 2016). Because of its character as a meta-process it is indeed difficult to say exactly where mediatization "begins" and where it "ends". Mediatization sceptics raise an important point when they argue that the term is too often applied in a very loose manner and ultimately "has no outside" (Deacon and Stanyer, 2015: 657); it may seem as though any kind of media-related change could be subsumed under mediatization. This problem becomes particularly obvious in social-constructivist analyses of mediatized social worlds where it is difficult to set up measurable indicators of how, for example, everyday activities are adapted to media. As mediatization researchers we need to be careful in how and when we attribute the term mediatization, ensuring that we maintain clear definitions of the sub-processes we study and a clear idea of how they relate to mediatization at large.

So what is it in Ruben's life story that makes it relevant to use as an example of mediatization? It would certainly not be enough to say that the mere fact that his life environment is crowded with media justifies a "mediatization diagnosis". Instead, what I want to advance in this book is *a cultural materialist perspective* that draws our attention to the ways in which media become *indispensable* to people's lives (see Chapter 2). Following the cultural materialism of Williams (e.g., 1974, 1977), I argue that this happens when media are thoroughly and commonly incorporated as *cultural forms* and it becomes difficult to imagine a life without them. As we saw in Ruben's case, it also happens when media technologies, along with certain ways of using them, are turned into normalized parts of the *doxa* of social fields (see Bourdieu, 1972/1977; 1997/2000), and thus raise the bar as to which modes of activity are available, even thinkable, to social agents and which are not.

Third, and related to the previous point, there has been a tendency in much mediatization research to not fully situate arguments in relation to *empirical contexts* and not provide convincing evidence of how particular cases and/or sub-processes are related to the overarching framework of mediatization. On the one hand, there is a risk of generating grand theoretical constructs that use a variety of studies as illustrations or examples in a more or less superficial manner (Ekström et al., 2016). On the other hand, there is a risk of drawing general conclusions regarding the long-term dynamics of mediatization based on limited empirical evidence. In this book I want to apply the above-mentioned cultural materialist approach to a set of empirical cases that, because of their social composition, allow for substantial analyses of how mediatization works on a general level as well as in relation to specific social, cultural and economic conditions. The empirical analyses are based on a series of qualitative research projects conducted in Swedish/Scandinavian contexts since 2003 and deal with three thematic areas: privileged expatriate lifestyles (Chapter 5), middle-class-biased cultures of urban exploration (Chapter 6), and urban and provincial gentrification (Chapter 7).

The analyses have three things in common. First, they are demographically linked to what we may broadly classify as *the middle classes*. The underlying reason for this choice is that the middle classes play a normalizing role in society, not least when it comes to consolidating certain ideals of connectivity and mobility, which in turn reinforce the mediatization process. Accordingly, the analyses of this book rest on the assumption that the middle classes provide a particularly fertile ground for grasping the dialectic of mediatization (see Chapter 4). Second, the analyses deal with various forms of *geo-social mobility* and spatial appropriation. Again, this is a choice made in order to analyse mediatization processes in contexts where we can expect to find accentuated tensions between individual autonomy and growing media dependence, between lived experience and the dominant ideals of our individualized, mobile culture. As Elliot and Urry (2010: 27-28) argue, "what is at stake in the deployment of communication technologies in mobile lives […] is not simply an increased digitization of social relationships, but a broad and extensive change in how emotions are contained (stored, deposited, retrieved) and thus a restructuring of identity more generally". Mobile lives are thus seen not just as a social characteristic, but represent what Williams (1977) calls a *structure of feeling* (see Chapter 2). Third, the analyses of this book pay considerable attention to individual and collective *biographies and social trajectories*. The reason is that mediatization *evolves over time*. We cannot draw any substantial conclusions about mediatization unless we compare current social conditions with conditions of the past in some way. One way of doing this is to look into the personal narratives and lived experiences of individuals like Ruben. Such individuals and groups are at the same time the active agents and the *reactive* elements, sometimes even the victims, of mediatization. In sum, at the empirical level this book provides critical, cultural materialist analyses of how mediatization processes are played out, experienced and culturally legitimized in relation to the social trajectories of mobile fractions of the middle classes.

This brings me to the fourth, and final, shortcoming of mediatization research that I want to address in this book. As stated above, mediatization is commonly understood as a meta-process, similar to globalization and individualization. It is also often stated that these meta-processes overlap and interact in complex ways (e.g., Hepp et al., 2015; Krotz, 2014). Nevertheless, there is still a lack of systematic assessments of how the relationships between these meta-processes are to be conceptualized and turned into empirically accessible areas of study (see Lunt and Livingstone, 2016: 462). The focus on privileged mobile lives is to be seen as a response to this situation. Through analyses of life biographies such as Ruben's we can develop *a critical bottom-up perspective* of how mediatization interacts with globalization and individualization processes, and ultimately delineate some of the key characteristics of this phenomenon (Chapter 8). How and why do media become indispensable within the lifestyles of mobile middle-class fractions? How are such mediatization processes interlinked with the quest for social autonomy and recognition and with the construction of privilege, status and power?

In the remainder of this introduction I further describe and contextualize the research problem of the book. This means, first of all, that I motivate and position the theoretical approach, and the book as a whole, in relation to the broader field of mediatization research. Second, I give an overview of previous research that has advanced critical and empirically grounded views of how mediatization processes influence contemporary identities and power relations. Third, I present and discuss the empirical data on which I build my analyses, aiming to further substantiate the value of studying mobile middle-class groups within a mediatization framework. At the end of the chapter I present the structure of the book.

The dialectic of mediatization

In recent debates around mediatization we have witnessed the consolidation of two main perspectives: the institutionalist and the social-constructivist (Couldry and Hepp, 2013). The institutionalist approach is concerned with how certain institutions in society, such as politics (e.g., Esser and Strömbäck, 2014), religion (e.g., Hjarvard, 2014) and corporate business (e.g., Ihlen and Pallas, 2014), adapt to the logics of dominant media institutions, notably mainstream journalism and commercial broadcasting. The social constructivist (or "cultural"; see Lundby, 2014a) approach focuses on how media and their representations, understood in a more complex sense, play into the ongoing construction of social worlds (Hepp, 2009; Krotz, 2007). In addition to these perspectives, Lundby (2014a) argues that one can also discern a third approach, which he calls "material". In the introductory chapter to *Mediatization of Communication* he describes the material perspective as having been influenced by the media ecological theories developed by McLuhan and the Toronto School, a perspective, it is implied, in which media are analysed mainly as material resources whose affordances, or biases, give opportunities for, or set limits to, various forms of agency (see also Clark, 2009). Furthermore, Bolin (2014) argues in another chapter in the same book that there is a "technological" perspective of mediatization, which he associates with the postmodern legacies of Baudrillard.

However, there is an important difference between these latter notions and the social-constructivist and institutionalist perspectives; that is, there are actually no contemporary researchers who actively propagate "material" and "technological" versions of mediatization research. Furthermore, the association of "material" with McLuhan and "technological" with Baudrillard seems to suggest rather incompatible strands of research, even though they point to a shared problem, namely the techno-material force of media development. The perspective that Lundby and Bolin identify should rather be understood as intersecting areas of media studies that address questions that could, or rather *should*, be incorporated within a mediatization framework. The perspective I advance in this book entails a broad concern with both materiality and technology, but it is quite different from the above-mentioned approaches in that it builds upon a cultural materialist tradition and is explicitly critical.

As I will discuss in Chapter 2, the critical potential of the mediatization framework has not been sufficiently elaborated in the debate so far. In my view, the critical potential is what makes mediatization such an important contribution not just to media theory, but also to cultural and social theory at large. Through mediatization we can reflect upon the ambiguous social and cultural implications of a media-saturated society; that is, what it is to live with media as normalized parts of the environment. Conceiving of mediatization as an essentially critical concept is to recognize how social processes, in a broad variety of domains, and at different levels, become *inseparable* from and ultimately *dependent* on processes and resources of technological mediation, and to identify the *feelings* and *experiences* that such dependences evoke. Speaking of dependence, in turn, is a way of ensuring that mediatization refers to something more specific than simply the increasing use or saturation of media in various realms of society or the quantitative growth in the circulation of data and information. Mediatization highlights *qualitative* transformations of socio-material relations – driven by social, cultural, economic and technological forces – whereby "media enabled" increases in the human capacity for social and cultural agency also incorporate decreases in individual or institutional autonomy. This, in turn, means that mediatization necessarily implies *a state of growing contradiction* that ultimately boils down to the tension between autonomy and dependence (see also Hjarvard, 2008: 17).

What do I mean when I say that mediatization always implies a negotiation of autonomy? One can of course argue, in line with the medium theorists, that media (understood as technologies and institutions) in many ways *extend* the capabilities of agents and institutions (see also Schulz, 2004), thus contributing to the strengthening of their autonomy. The Internet has made it possible for anyone to search for potentially liberating information to an extent that is historically unprecedented. Mobile devices have lowered the threshold for many people to become mobile, to feel safe and secure when they move about in familiar and unfamiliar environments. But at the same time the growing reliance on media to keep informed, stay in touch and carry out various transactions ties individuals and institutions closer to the technological infrastructure and institutional logic of media. This is an example of what Giddens (1991) refers to when he discusses the growing necessity of trust in abstract systems.

However, negotiations of autonomy do not only occur in relation to various *orders of technology*. There are also cultural and social *orders of recognition* that feed into these processes (Figure 1.1). Technologies do not become indispensable in and by themselves, as an effect of intrinsic affordances, but when they enter into social relationships that give them a certain meaning and place. This is particularly important to keep in mind when we study media because their very objective, in contrast to other technologies, is to sustain communication between people. For

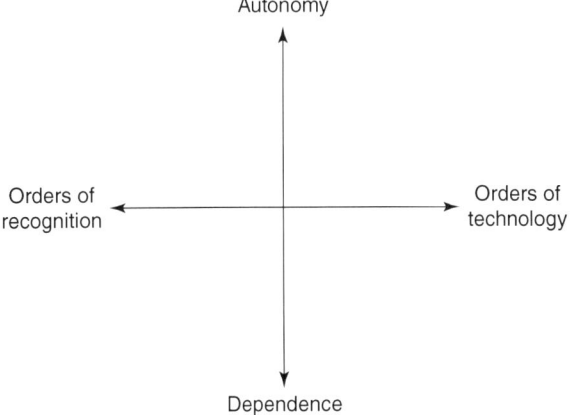

FIGURE 1.1 The dialectic of mediatization.

instance, let us think of the gradually normalized role of Facebook as part of our common culture. During the last decade many people have come to experience Facebook not merely as a superficial pastime but as an everyday necessity for staying in touch with friends and family and for keeping abreast with what is going on in their social environment. While the user-friendliness of mobile devices and digital interfaces may lubricate this process of socio-technological integration and normalization, we should also consider how such developments are grounded in the social need for recognition. In the context of media, technological dependence cannot be understood in isolation from various orders of recognition – orders that are the pre-condition for individual growth and autonomy while at the same time potentially invoking social restraints. These tensions and dynamics, as illustrated in Figure 1.1, underpin the dialectic of mediatization that is at the centre of this book.

Accordingly, my focus on *recognition* is integral to the dialectical approach. As given by Hegel's phenomenology, the fact that individuals *need* recognition from others in order to achieve a sense of selfhood and social worth implies that recognition is a dialectical construct from the outset. But while recognition is to be seen as a fundamental element of social life and identity development it cannot be taken for granted. *It can fail*, especially if acts of recognition are experienced as not genuinely intended or as parts of an asymmetric relationship (see Hegel, 1977). This vulnerability has become more obvious under conditions of increasing individualization in which recognition has to be achieved through reflexive forms of adaptation and connectedness. People today can still, to a lesser extent, rely on fixed social structures but are forced to put more work into the creation of themselves as both autonomous *and* socially and culturally recognized subjects. If successful, these processes can lead in emancipatory directions. But they may also reproduce social anxieties and experiences of restraint. Here I build especially on Honneth's (2004, 2012) social theory of recognition and his critical writings on *organized self-realization*. The latter concept points to structural transformations whereby personal development has become intertwined with consumerist forces of capitalist society and thus normalized according to standardized patterns of classification. I also build on Bourdieu's (1979/1984) work on lifestyles and cultural distinction, which asserts that processes of cultural recognition and misrecognition are integral to the continual struggle for social esteem and prestige. As McBride (2013: 83) argues in his overview of recognition theory, if we locate processes of recognition in a contested space of social positions, "we can see that there is a continuity between the personal, emotional, lives of individuals and the larger, institutional, context in which these lives are lived".

This is precisely the type of continuity I want to illuminate in this book. At the most mundane level of analysis, media are used as a means of connectedness and are thus integral to, and legitimized by, everyday processes of recognition. Behind these processes, however, there may also exist feelings of stress, frustration, emptiness and restraint stemming from the discrepancy between what media are expected (and often proclaimed) to deliver and what gratifications are actually achieved. At a more structural level, media practices are themselves classified and are thus part of the ongoing structuration of society whereby certain groups and their lifestyles are recognized as more prestigious than others. Furthermore, media as such constitute institutions for the circulation of myths and ideologies concerning which forms of recognition are to be seen as desirable among different groups and potentially beneficial to their quest for autonomy and self-realization. Whether these avenues to positive recognition are accessible or not, however, remains a critical matter that ultimately points to the stratification of mediatization processes.

In short, the uneasy relationship between orders of recognition and orders of (media) technology generates a series of contradictions that together set the agenda for critical mediatization research. In its focus on such inherent contradictions of modern society, I argue, mediatization research should follow a tradition of *immanent critique*. As Fornäs (2013: 510) points out, this is a powerful form of critique that strives to unveil the inner ambiguities of dominant relationships in order to put those "contradictions and tensions in movement, rather than freezing them to a standstill". Immanent dialectical criticism is distinct from transcendental criticism, or what Walzer (1987) calls "inventive" criticism, that rather "raises an external ideal image against the prevailing social and cultural conditions" (Fornäs, 2013: 509). While mediatization is not a normative concept, there is an inbuilt transgressive potential in mediatization research based on the fact that ambiguities and contradictions integrate the prospects for further change. Accordingly, critical mediatization research does not have to point out the particular directions in which such changes ought to lead. The analytical endeavour is to expose how various dependences are shaped, how they contradict "socially incorporated ideals" (Honneth, 2009: 53), and what their consequences are. Such analyses *per se* have important social implications.

Current transformations of media society

This book is an attempt to say something about the directions in which mediatization pushes contemporary culture. To some extent the analyses are even prognostic, in the sense that they focus on groups, lifestyles and processes that are understood to be at the "forefront" of mediatization. This may seem a problematic ambition given the above-mentioned view of mediatization as a long-term meta-process of social change. Is it at all possible to speak about mediatization without carrying out longitudinal research within certain social contexts? There are of course limitations to what we can learn from the type of interview studies that a major part of this book is based upon (see Chapters 5–7). They can provide only glimpses of the broader trends that mark mediatization. However, if we want to gain deeper understandings of what it feels like to live in a mediatized culture and how cultural and material normalizations come about within everyday practice we must talk to people about their lives and their *experiences of continuity and change*. Through cross-projections of such personal accounts we may grasp the dialectic of mediatization as well as changes in culturally and materially shaped lives, what Williams (1977) terms *structures of feeling* (see Chapter 2).

It is also important to work cumulatively. Researchers have already produced a relatively complex picture of contemporary mediatization processes, even though it is far from congruent in all respects, and even though the very term mediatization has not always been used. There are thus a number of empirical findings and conceptual constructs that can provide a context for the dialectical problem of this book. In the following two sections I briefly overview this context. First, I advance *transmedia* and *polymedia* as complementary markers of today's media landscape and discuss how they are related to an expanding *culture of connectivity*. Second, I highlight a selection of studies that have in various ways addressed the ambiguous sociocultural *consequences* of this development.

Transmedia, polymedia and the culture of connectivity

If we are to understand the current consequences of mediatization at the level of social life – that is, how various realms of activity, such as social fields (Bourdieu, 1980/1990), become adapted to and dependent on different technologies and institutions of mediation (Jansson, 2013a) – we need to tease out what is qualitatively new about the current media situation. This is not to say that long-term social processes and pre-established media structures are of less importance, but we should try to identify the new directions in which mediatization drives social change. In this regard, a number of more or less totalizing concepts and metaphors have been suggested along with the expansion of digital media, ranging from Castells's (1996) "network society" to Jenkins's (2006) "convergence culture", from Poster's (1995) "second media age" to Deuze's (2011) "media life", just to mention a few examples. If we want to disentangle the cultural–material conditions of mediatization, however, we need more specific terms.

At the technological level, we can speak of an ongoing transformation from mass media to *transmedia technologies*. The term transmedia was originally introduced by Jenkins (2006) in his analysis of emerging forms of "transmedia storytelling" enabled by the interlinking of different digital technologies and platforms. In broader terms, transmedia points to a condition in which content flows between, and is constituted by means of, a multiplicity of channels. This goes for concepts generated by culture industries as well as for the everyday circulation of private information. Contemporary Hollywood films, for instance, are produced at the same time as complex "brand-worlds" and as "story-worlds" in which narratives and promotional material not only *include* different channels, but *flow across* channels, and in which fans and other audience members become agents of the symbolic circulation (Fast, 2015). Similarly, transmedia technologies have made it possible for ordinary people to access, circulate and share their private files (whether text, sound or image) and consume a range of media flows (e.g., TV programmes, social media updates and video clips) via different platforms and devices. The "spreadability" of content, its capacity to spread through the sharing practices of various audiences, has become a key asset within informal as well as formal online economies (Jenkins et al., 2013). As I will discuss in Chapter 3, such transmedia affordances are increasingly taken for granted.

One thing that distinguishes our contemporary media landscape from media landscapes of just one or two decades ago is the *social location* of media. In addition to their "traditional" position between people and various organizational entities (including media institutions) that characterized the mass media landscape (see Hjarvard, 2013: 23-27), media technologies are now to a greater extent located *between people*. This is not to say that interpersonal media are all new; telephony and the postal system have been crucial to the history of modernity. Nor is it to say that today's transmedia technologies have replaced mass media. Rather, these forms co-exist and interact in various ways, giving rise to complex media landscapes.

Madianou and Miller (2012) introduce *polymedia* as a concept for explicating how this growing complexity influences social life. Polymedia thus takes us beyond the strictly technological realm of transmedia. It refers to a socio-material condition as well as to an epistemological approach to media practices. Regarding the first aspect, polymedia denotes an everyday media environment in which individuals can manage their social relationships through a variety of interchangeable devices and applications. Under such "media abundant" conditions, the decision as to which media to use for making contact does not involve any economic considerations, but is dependent rather on which device and application are felt to be most adequate in moral and emotional terms for a particular type of relationship, content and/or situation. Madianou and Miller present a clear definition of the criteria by which one may talk about polymedia in the strict sense of the term. First, the individual must have access to "a wide range of at least half a dozen communication media" (ibid.: 126). The authors do not justify why half a dozen should be the limit, but the main point is people's ability to afford multiple hardware items (such as both a mobile phone and a computer). Second, the individual has to be sufficiently media-literate to make use of media effectively. Finally, polymedia implies that the costs of communication are linked to infrastructure (especially subscriptions of different kinds) rather than to the individual act of communication. For example, the price of sending fifty e-mails or text messages is the same as the price of sending one. In sum, polymedia is not an evenly distributed phenomenon. It implies a material state that can only emerge under conditions of relative affluence. Nevertheless, it has gradually become the normalized state of living for millions of people in contemporary (post-)industrialized societies.

In epistemological terms, polymedia takes its cue from Levi-Strauss's (1963) anthropological structuralism, which actualizes an approach to media use that treats the media as a symbolic environment rather than as individual channels of communication: "Polymedia is a form of structuralism in which the understanding we have of any one medium becomes less its properties, or affordances, and more its alternative status as against the other media that could equally be employed for that message" (Madianou and Miller, 2012: 137). This approach also has a great deal in common with Bourdieu's (1979/1984) cultural sociology and brings about a renewed interest in cultural distinctions related to interpersonal media practices.

Madianou and Miller note that distinctions were established in earlier stages of media development as well; for example, they point to the significance of letter writing as a sign of educational achievement among Filipino migrants in the United Kingdom (ibid.: 57). However, the increasing number of technologies and applications that individuals can use for communicating with one another today requires that the sign value, as well as the social functionality, of each medium is re-negotiated in relation to a more complex and rapidly shifting media environment (see also, e.g., Baym, 2010; Broadbent, 2011). It becomes increasingly difficult to define to what extent classified preferences and distinctions, as manifested through everyday cultural practice, pertain to certain devices (smartphone, laptop, etc.), applications (Skype, WhatsApp, etc.) or modalities (text-based, audiovisual, etc.) of communication.

The rise of transmedia and polymedia thus translates into a state of blurred distinctions and open-ended processes of mediation. In a similar vein, Deuze (2011, 2012) argues that people in affluent, networked societies no longer live *with* media but *in* media. This means that media are not external to everyday practices; rather, a continuously expanding array of practices that were previously carried out without media or only supported by specific technologies are now carried out *within* the realm of online media. This goes for a variety of consumption practices as well as for social interactions such as dating and sharing of experiences. As noted above, Deuze calls this new condition "media life", which, according to him, expands as media users become increasingly involved in the production and circulation of mediated content, notably via social media platforms. Such media may take the form of websites or mobile applications and involve social networking (e.g., Facebook, LinkedIn), video sharing (e.g., YouTube), blogs and microblogs (e.g., Twitter, Weibo) and social media extensions of various lifestyle applications (e.g., RunKeeper, Nike+). What these media have in common, besides the fact that they link people together, is that they turn social practices into economic value through the automatic aggregation and algorithmic processing of user-generated data (see also, e.g., Gillespie, 2010; Van Dijck, 2012, 2013; Striphas, 2015).

Deuze's (2012) influential notion of "media life" captures the social relocation of media in an accessible way, but it tends to generate more questions than it answers. In fact, Deuze provides a rather sweeping view of social transformations, underplaying resistance as well as socio-cultural variations, and he fails to suggest any clear-cut sub-categories or specifications for how "media life" should be defined. If we are to build a situated critique of contemporary mediatization we need more precise and analytically useful concepts as well as empirical accounts that pinpoint social change. Here, Van Dijck's (2013) outline of an expanding *culture of connectivity* stands out as a particularly well-argued diagnosis, which also resonates with sociological theories of identity and recognition in individualized societies. According to Van Dijck (2013: 3–4), the rise of commercial social media has led to a rapid shift "from networked communication to 'platformed' sociality, and from a participatory culture to a culture of connectivity". It is important to distinguish *connectivity* from *connectedness*. Whereas connectedness refers to the meaningful social connections between individuals and groups – which social media promote and which various media have enhanced and extended in different ways since their inception – connectivity refers to "the socio-technological affordance of networked platforms to connect content to user activities and advertisers" (Van Dijck and Poell, 2013: 8). This means that the social practices that these platforms mediate are actually not as free and open-ended as one might think, but are partly governed and exploited via the algorithms of the techno-economic architecture. This is also why Van Dijck (2013) uses the term *connective media* instead of social media. In everyday life the distinction between connectivity and connectedness can be difficult to identify since many close relationships are also exploited and reproduced via automated connective processes. The important point, which I will return to in Chapter 4, is precisely this accentuated fuzziness between connectedness and connectivity, the fact that orders of recognition become pre-mediated and simulated through automated patterns of connectivity.

The culture of connectivity thus provides a straightforward entry-point to the dialectic of mediatization. It actualizes how desires for recognition are automatically exploited and reproduced, invoking even more complex social entanglements with dominant orders of technology. Adolf (2014) describes this development as "involuntary mediatization" in an attempt to pinpoint that people are regularly unaware of the extent to which their social interactions and movements in physical and online spaces are monitored, datafied and increasingly dependent on new information and communication technologies. Algorithmically processed "big data" can be turned into personalized prescriptions regarding anything from medical advice to musical tastes, and cultural intermediaries are replaced by "infomediaries" (see Morris, 2015). Similarly, Ritzer uses the term "prosuming machines" for stressing how connective and prosumptive activities on behalf of media users are no longer a matter of free choice but an automatically accomplished condition whose utmost expression comes with the "Internet of things" (see also Ritzer, 2015). Bakardjieva and Gaden (2012), in turn, advance Foucault's notion of "technologies of the self" as a way forward for understanding the interplay between liberation and domination involved in Web 2.0 technologies.

From a broader perspective, the culture of connectivity legitimates and is legitimized by what Dean (2009: 2) calls *communicative capitalism*: "the materialization of ideas of inclusion and participation in information, entertainment, and communication technologies in ways that capture resistance and intensify global capitalism". Participation and individual expression are no longer seen as civic privileges and sources of liberation, but as ideologically imposed demands that reproduce the system itself. The same hegemonic force is detected in Mejias's (2013: 9) notion of *nodocentrism*, suggesting that "what we are seeing is not only the pervasive application of the network as a model or template for organizing society but also the emergence of the network as an episteme, a system for organizing knowledge about the world". This means that the whole idea of networks and *networking*, largely promoted by global capitalism, takes over people's worldviews, marginalizing other ways of seeing the world as well as places and agents that do not conform (deliberately or due to lack of resources) with the vision of a society built around networked interaction and a "politics of visibility" (see Hillis, 2009). At the concrete level, nodocentrism is reproduced through, for instance, algorithmically generated recommendation lists (of friends, services, goods, and so forth), search engine results and automated estimations of security threats based on dataveillance.

Taken together, these tendencies give rise to ethical and political considerations concerning the status of free will and social autonomy in mediatized societies. Connectivity and networking may well be promoted discursively as the gateway to social recognition and integration, and ultimately autonomy, but they also bring about dilemmas of inequality (between those included and excluded), dependence (among those whose lives and careers are sustained through connective practices) and, as we shall see, social and existential discomforts.

The discomforts of mediatization

The contemporary ambiguities of individual autonomy and privacy have been intensely debated within media studies, surveillance studies, sociology and the wider terrain of the social sciences (see, e.g., Baym and Boyd, 2012; Fuchs, 2014; Andrejevic, 2013). In his much-cited book *The Filter Bubble*, Internet activist Eli Pariser (2011) sketches the downsides of algorithmically generated information streams that appear as personalized search engine results or "news flows" on social media. This practice, which Andrejevic (2007) terms "digital enclosure", manipulates media users to reproduce pre-existing values and interests and become socially trapped within enclaves of like-minded. Along the same lines, Lovink (2011: 34) argues that we must overcome "the paradoxical era of individualization that results precisely in the algorithmic outsourcing of the self". According to Lovink, in the current stage of mediatization (even though he does not use that term) the Internet has become so naturalized that it eventually disappears from our awareness, but gains "even more power on the level of the collective unconscious" (ibid.: 35). As a response, people in general should become more aware of the architecture of the filters that surround them. Furthermore, instead of simply going offline, a more conscious and radically experimental politics should be developed, according to Lovink (ibid.: 164):

> In order to be open to radically different possibilities, we need to say farewell to the "trust" paradigm that conceptually supports paranoid security systems and culminates in "walled gardens". The "risk" discourse should no longer only apply to entrepreneurs who are praised for their courageous risk-taking (with other people's money) while the vast majority of users remain locked into "trust" cages. Networks should not only replicate old ties. They have another potential. We need to abandon the "friends" logic and start to play with the notion of dangerous design.

Lovink's observations underscore the dialectical nature of mediatization. While further connectivity and information access may be a liberating opportunity for many, growing numbers of people suffer from information overload. The continuous handling of networked information is increasingly turned into (unpaid) work, affecting above all the middle classes who fear stagnation if they do not comply with the general doxa of organized self-realization (see Honneth, 2004). Social media also, expanding under the promise of providing connectedness and genuine recognition between people, demand continuous work that ties people even closer to corporate orders of technology and reproduces segregated patterns of interaction.

It is thus important to study the various social and existential discontents that mediatization actualizes among relatively privileged groups in relation to their social trajectories. This is not to say that these "problems of the privileged" are more important than the exploitation and general lack of resources that many other groups and indeed entire populations suffer from. But if we can grasp the ambiguities among those who aspire for status, I argue, we have a better chance of understanding the complexity of mediatization at large.

A case in point here is the transformation of white-collar work. As Gregg (2008, 2011) argues, the ubiquitous nature of networked, mobile media contributes to the normalization of "flexible work". Mid-rank professionals and various kinds of "knowledge workers" are enabled to invest more time and energy in their careers while at the same time being able and *expected to* entertain family life and close relationships. This new flexibility has often been celebrated as emancipatory for women. But it also incorporates considerable elements of stress and anxiety due to the fact that intimate relations have to be either managed at a distance or placed in the background of professional tasks. Furthermore, the prevalence of mediated "recognition work", which I discuss further in Chapter 4, is closely linked to the triumph of self-monitoring and social networking as normalized parts of working life. As Gregg's empirical studies show, competitive professionals often let work dominate their intimate relationships and spaces while certain aspects of their private selves are monitored in order to strengthen their professional identities – sometimes shaped publicly as personal brands.

Similar conclusions are drawn by Marwick (2012, 2013), who sees the ongoing institutionalization of social media for communication and self-promotion within many professions and branches as intimately tied to an expanding regime of social surveillance (see also Jansson, 2015b). At the same time, however, Marwick (2013: Chapter 4) found in her study of technology workers in San Francisco that social media, accompanied by the neoliberal discourse of self-branding and the "entrepreneurial self", were an important resource for those who wanted to reject the organizational imperatives of their employers and start up their own businesses in consulting or freelance.

These studies illustrate the interplay between orders of recognition and orders of technology whereby mediatization expands into various realms of activity. They also contribute to the grounding of our research problem in the actual settings of everyday life, clarifying how emancipatory efforts are often accompanied by experiences of discomfort. Even more concrete examples can be found in studies of how people manage and experience their media use on a day-to-day basis. As mentioned above, transmedia have a tendency to drag people into open-ended circuits of prosumption; polymedia represent environments where there are sometimes too many options to handle; and the culture of connectivity normalizes expectations of staying online, connected and tuned in most of the time. Such conditions may easily give rise to contradictory experiences of time wasted, opportunities missed and filters abolished.

The general argument is that mediatization under these conditions tends to bring about a sense of *simultaneously gaining and losing control*. As Bengtsson (2015a) points out, based on a qualitative interview study, a common experience among ordinary people today is the feeling that they need to find strategies to *disconnect*, at least temporarily, from continuous information (over)flows. Among her interviewees there are people who deliberately leave their mobiles at home when they go for a walk, or try to discipline themselves not to bring their mobile devices into the bedroom. Such acts of resistance can be seen as evidence of the socially negotiated force of mediatization. They are also evidence of the uninterrupted relevance of moral structures that prescribe what levels of media use are to be understood as appropriate, governing when feelings of *having used media too much* set in (see also Hall and Baym, 2012; Bengtsson, 2011). Ultimately, such structures are articulated through various actions of *non-use* or *disconnection*. As well as the fact that there are groups in society that generally resist, or at least try to minimize, the penetration of digital media in their everyday lives for ideological or cultural reasons – which can sometimes be interpreted as a sign of socio-cultural privilege and independence (Kaun, 2014; Kaun and Schwarzenegger, 2014; Jansson and Lindell, 2015) – there are also examples of more outspoken "anti-mediatization" movements. They may target, for instance, the cultural and commercial dominance of social media platforms like Facebook (e.g., Portwood-Stacer, 2013) or the perceived social pollution associated with extended smartphone use at live events and other public gatherings (e.g., Hutchins, 2016). Further examples can be found among artistic movements that actively attempt to disrupt the prevalent functions of media in order to problematize their role in society (e.g., Pinder, 2013). In my view, such forms of resistance, whether explicitly political or anchored in the realm of everyday life, are to be seen as *reactions* to mediatization; that is, as cultural symptoms of the dominant directions that media-induced social transformations are currently taking. I return to this question in Chapter 8 where I discuss the prospects of *counter-mediatization*.

On the other side of the coin, experiences of *involuntarily* losing connection can be very unsettling. Paasonen (2015) gathered autobiographical reflections from her university students in order to trace the feelings evoked by disconnection caused by failing networks. She found that the essays diverged from the dominant understanding (within certain research strands) of media use as a form of rational agency. Instead, the essays depicted user agency as "ambivalent connectivity to and dependency on various networks" (ibid.: 703). Above all, the students told stories about recurring frustrations with malfunctioning technologies and networks, involving strong sensations of annoyance with the fact that certain things cannot be fixed immediately and that time is thus wasted on just waiting. Some informants had even decided to use simpler technologies in order to minimize the risk of problems.

It should be stated here, of course, that experiences of information overload, problems of maintaining the boundaries of privacy, or feelings of guilt and shame pertaining to one's excessive or otherwise immoral media use, are not at all new. They have not emerged as a consequence of transmedia technologies or polymedia environments, or as part of the hegemonic culture of connectivity. Rather, they are to be seen as general expressions of the dialectic of mediatization. There are a number of studies from the 1980s, 1990s and early 2000s demonstrating how people had to struggle to develop everyday strategies in the era of broadcasting as well in order to defend their private spaces from various mediated intrusions and maintain a sense of moral balance regarding their media use. As a case in point, we may recall Hirsch's (1992) study of a British middle-class family and their rejection of the video-recorder because of the perceived risk of watching too much television instead of spending time on more "legitimate" activities (see also Morley, 1992, 2000; Jansson, 2003a).

Still, it is obvious that today's media environments are *denser*, more *complex* and qualitatively *different* from the ones people were once used to. At a time when media saturate most realms of activity, social and technological dependences and adaptations must evolve in new ways (Jansson, 2013a). Similarly, *new forms of discipline* are required of individual media users who wish to stay in control of the time-spaces of work and leisure, public exposure and intimate relations (see, e.g., Boyd and Marwick, 2011; Christensen and Jansson, 2011; Jansson, 2014). These are changes that I discuss further in Chapters 3–4, as well as in the analyses of Chapters 5–7.

Analysing mobile lives

This book provides not just a critical account of mediatization; it is also an exploration of mobile middle-class lifestyles and life trajectories. The term *middle class* is applied here as a relational construct, following the theorizations of Bourdieu (1979/1984). My point is not to objectively compare the media uses or preferences of particular socioeconomic groups, but to assess what it means to handle the dialectic of mediatization under life-conditions marked by *relative affluence* and normative *expectations of social mobility*. I thus refrain from delimiting middle-class life merely according to socioeconomic criteria (even though such factors are of course important for the general shaping of these social positions), but rather approach it as an *aspirational social space* (see Polson, 2016: 8-9) where social agents are required to continually negotiate their cultural and material life-conditions as a means of improving their social status. The middle classes are to a great extent the outcome of modern social transformations whereby new occupations, new life-environments and new forms of social practice have emerged and replaced older structures. While their precise levels of education and income, or the balance between cultural and economic capital, may vary, what unites the middle classes (across national boundaries as well) is that their inherited pre-dispositions do not easily fit the social positions and lifestyles that are classified as desirable.

My focus on the middle classes is motivated precisely by their intermediary role within mediatization processes, the fact that their lifestyles and life trajectories are founded upon a relatively open-ended *habitus* (see Bourdieu, 1979/1984: chapter 3). They are, on the one hand, key agents of mediatization because of their relative affluence and desire to keep abreast of ongoing developments. They even maintain a hegemonic role in the sense that their ways of life (speaking in general terms) are seen as the norm in most individualized democratic societies. The middle classes can be seen as the champions of organized self-realization (Honneth, 2004), by which is understood an individualistic order of recognition. On the other hand, their positions are also vulnerable and unstable precisely because they are "in-between", which means that middle-class lifestyles are often marked by sensitivity to new trends and exposed to various degrees of anxiety. It is among these groups, I argue, that we can most readily grasp the general ambiguities of mediatization.

My focus on *mobile* middle-class lifestyles is linked to the fact that mobility is often seen as a sign of success, even a social norm, tied to middle-class life. As Cresswell (2006) argues, we live in times dominated by the metaphysics of flow, meaning that mobility is culturally associated with progress and flexibility. Being mobile, or, rather, being in control of mobility, is thus a desired condition, especially for those aspiring to status positions within (post-)industrial economies. Mobility has become a norm that dominates working life and leisure interests as well as entire life-biographies (see also Elliot and Urry, 2010; Polson, 2016). In a similar vein, Thrift (1996: Chapter 7) argues that modern lives are marked by an accentuated *mobility structure of feeling* shaped by imaginative *adjustments* to the "machinic complexes" of speed, light and power (see also Urry, 2007: 6–7). This approach corresponds to the dialectical understanding of mediatization (see Figure 1.1) and points to the ways in which mediatization is related to the normalization of mobile lives. Staying in control of mobility means that the individual must also learn how to handle media, and more precisely *connectivity*, in efficient and culturally appropriate ways, which in turn implies that he or she is exposed to the discomforts of stress, frustration and anxiety, as discussed above.

I should emphasize here that my focus on mobile lives is not the same thing as a focus on mobile media. While the general expansion of mobile media is certainly an important aspect of contemporary mediatization processes, my analyses aim to grasp a broader picture, including all kinds of media and their combined cultural-material significance. I see this book as a contribution to a social theory of the relationship between mediatization, individualization and globalization rather than as an inquiry into the role of particular media. As such, the book is closely aligned with the so-called "new mobilities" paradigm in the social sciences, "putting social relations into travel and connecting different forms of transport with complex patterns of social experience conducted through communications at-a-distance" (Sheller and Urry, 2006: 208).

Part I of this book sets out to further elaborate the cultural-materialist approach (see especially Chapters 2–3) and substantiate the analytical focus on mobile lives as a contemporary structure of feeling (following Williams, 1977; see Chapter 4). Part II, in turn, contains empirical analyses pertaining to three specific sites of this structure of feeling: *elite cosmopolitanism* associated with Scandinavian expatriates working for the United Nations (Chapter 5), *post-tourism* in the specific context of urban exploration (Chapter 6), and *gentrification* processes in urban and rural areas (Chapter 7). These studies provide deeper understandings of how mediatization unfolds in relation to different middle-class mobilities and thereby shapes the broader meanings of privilege and status. The analyses have two epistemological features in common. First, they are to be understood as *socially oriented* (Couldry, 2012) or "non-media-centric" (Morley, 2009; Krajina et al., 2014) in the sense that they grasp the role of media in a holistic way, approaching media as *cultural forms* embedded in particular ways of life (Williams, 1974) rather than as technologies or texts. All analyses are based on qualitative interviews that focus on everyday life and socio-cultural relations in broad terms so as to represent the role of media in a balanced way and to grasp mediatization in terms of socially and culturally moulded sub-processes rather than as one overarching force. A more synthesized view, in which the general directions and power relations of mediatization are addressed, is developed in the final chapter of the book (Chapter 8).

Second, the studies give considerable attention to the *life biographies and social trajectories* of individuals. The underlying idea is that mediatization can only be understood by identifying long-term social transformation. Such knowledge may seem to be at odds with the type of qualitative media sociology exercised in this book. However, as mentioned above, through the cross-projection of individual life stories we can achieve a complex as well as a historically meaningful picture of what has characterized mediatization during the last few decades. Even though this is a rather short time-span, it is a period of rapid technological change that harbours the entire shift from mass media dominance to expanding transmedia infrastructure, as well as major social changes tied to globalization and new forms of mobility.

So what kind of empirical material, more precisely, is this book based upon? During the last fifteen years I have directed and been involved in a number of projects dealing with the uses of media in various socio-cultural groups and environments. Findings have been reported in articles and books over the past decade, but for the most part separately. What I want to achieve with this book is a synthesis of material and findings gathered across different projects between 2003 and 2016 in order to generate a more comprehensive understanding of the social consequences of mediatization. The main empirical sources are the following 60 interviews (listed in chronological order):

1   Eight interviews conducted in 2003 with people moving into the new apartments of the Western Harbour area in Malmö, Sweden (see Chapter 7). The turn of the millennium marked the starting point for the huge urban regeneration of the former harbour area, signified by the 2001 *City of Tomorrow* international housing exhibition. The regeneration project was from the start saturated with the metaphysics of flow, promoting the new district as a node in the network economy. The interviews were conducted within the research project *Culture, Identity and Life Forms in the Post-Industrial City*, funded by the Swedish Research Council and directed by Professor Ove Sernhede, University

of Gothenburg. During this period I lived in Malmö and worked at Malmö University, which was then located in the Western Harbour and thus an embedded part of the post-industrial transformation.

2   Six interviews conducted in 2008 with Scandinavian expatriates living and working in Managua, Nicaragua, who were employed by various international development organizations (Chapter 5). These interviews were part of a non-funded project on media and cosmopolitanism I was working on during a period of research leave spent in Managua (five months in total).

3   Fourteen interviews conducted between 2009 and 2010 with people living, working and/or residing in Arvika, a countryside community in the county of Värmland, Sweden (Chapter 7). Among the interviewees there were a number of individuals who had moved *to the countryside* – four of them immigrants from the Netherlands – and who could thus be seen as counter-urbanizers. Their lifestyles were heavily dependent on well-functioning media infrastructure, while the countryside was also positioned as a retreat from information flows. The research was conducted within the research project *Rural Networking / Networking the Rural*, funded by the Swedish Research Council FORMAS and directed by me. I carried out all interviews except for the ones with Dutch immigrants, which were conducted by MA student Joice Tolentino. I also have long-established personal connections to the Arvika municipality.

4   Ten interviews conducted in 2013 with artists and cultural entrepreneurs in the municipality of Arvika, Sweden (Chapter 7). The respondents were active within the areas of theatre, music, film, literature and/or arts and crafts. All of them had *moved* or *moved back to* Arvika, which made them part of a broader trend of counter-urbanization and, potentially, provincial gentrification. The interviews were conducted by Dr Linda Ryan Bengtsson within the research project *Rural Networking / Networking the Rural* (see above).

5   Thirteen interviews conducted in 2014 with Scandinavian expatriates working for international organizations, mainly within the UN system, in Geneva, Switzerland (Chapters 5 and 7). These interviews were part of the research project *Kinetic Élites: The Mediatization of Social Belonging and Close Relationships among Mobile Class Fractions*, funded by the Swedish Research Council and directed by me. The project analysed mediatization processes within three social fields, and I was responsible for the field of international politics, development and diplomacy. During 2014 I spent two periods of fieldwork in Geneva (six weeks altogether).

6   Nine interviews conducted in 2015 and 2016 with people associated with the Swedish urban exploration community (Chapter 6). These interviews were carried out by PhD candidate Tindra Thor as part of the research project *Cosmopolitanism from the Margins*, funded by the Swedish Research Council and led by Professor Miyase Christensen, the Royal Institute of Technology, Stockholm. The material included "go-alongs" as well as interviews conducted in person or via Skype. Geographically, the interviews were spread out over different regions of Sweden.

The empirical material thus contains both interviews and first-hand ethnographic observations from each of the above-mentioned areas. This means that I have gained not only a multi-sited, but also a long-term understanding of the phenomenon – empirically as well as theoretically. In addition to these primary sources I draw on interviews conducted within the research project *Secure Spaces: Media, Consumption and Social Surveillance*, funded by the Swedish Research Council for the Humanities and Social Sciences (*Riksbankens Jubileumsfond*) and directed by me. These interviews were conducted by research assistants during 2011 and 2012 and included locally based middle-class dwellers of a Swedish small town (twelve interviews) and well-educated middle-class dwellers in the inner city of Stockholm (ten interviews). I do not address these sources in separate analyses/chapters of the present book, but refer to them occasionally, mainly in Chapter 3, in order to illustrate specific theoretical points. All interviewees figuring in this book have been anonymized.

Outline of the book

As already mentioned, this book is divided into two main parts. Part I (Chapters 2–4) constitutes a critical engagement with the expanding field of mediatization research and Part II (Chapters 5–7) presents empirically grounded analyses of mediatization processes in the context of mobile lives. Chapter 8 contains a concluding, synthesizing and forward-looking discussion.

Chapter 2 – Mediatization Is Ordinary – assesses the value of cultural materialism for developing a critical approach to mediatization. Such a critical approach can be seen primarily as a response to the difficulty of identifying *what is not* mediatization in the realm of culture and everyday life, a way of providing sharper contours to what we mean by mediatization. The argument is developed through a discussion of Raymond Williams's (e.g., 1974, 1977) work on cultural materialism, notably his concept of cultural form, along with Bourdieu's (e.g., 1972/1977) theory of social fields. Using these theories the chapter advances the intermediary terms *ordinary culture* and *communicational doxa* as the lenses through which to study mediatization as a cultural *and* a material force. Mediatization is outlined as a broad societal transformation whereby continuous everyday adaptations to, and negotiations of, media as socially amalgamated cultural forms go hand in hand with the (trans)formation of power relations.

Chapter 3 – Why Are Media Indispensable? – expands on the cultural materialist approach in relation to the media environments of everyday life. It addresses the question of how and why media technologies and related things become *indispensable*, and how places and practices become materially adapted to the existence of media. Indispensability is here understood as a bonding force between social agents, technologies and the world. The chapter introduces a triadic framework for the study of material indispensability and adaptation, corresponding to a triadic articulation of media as cultural forms. It includes Ihde's (1990) notion of "I-technology-world" relations (*media technics*), Bourdieu's (1979/1984) theories of socio-cultural legitimation and practical knowledge (*media properties*) and Lefebvre's

(1974/1991) phenomenology of the materialization of everyday life *(media textures)*. Taken together, these perspectives can help us identify different forms of media dependence as well as internal tensions and fluctuations within the mediatization meta-process as it unfolds in relation to different technological regimes, during different periods and in diverse socio-cultural contexts. The chapter pays particular attention to the ongoing shift from "mass media textures" to "transmedia textures", signifying the coming of a new sub-stage of mediatization: *transmediatization*.

Chapter 4 – Social Recognition and Status in a Mediatized World – links the cultural materialist perspective to an analysis of social class and the cultural articulation of status. The main purpose is to show how mediatization is historically linked to individualization and how the interaction between these meta-processes gives shape to dominant orders of recognition in contemporary middle-class settings. These discussions expand on Honneth's (e.g., 2004) critical theories of recognition and *organized self-realization* and Bourdieu's (1979/1984) analyses of how *social status* is (re)produced through ongoing struggles over symbolic resources. Mutual recognition is a pre-condition for gaining social autonomy while at the same time demanding much reflexive work, especially in socially dynamic contexts. The final part of the chapter discusses how the quest for recognition is articulated among the mobile middle classes, whose normalization of certain media properties plays an ambiguous role in relation to the hegemonic structures of mediatization.

Chapter 5 – Mediatization and Elite Cosmopolitanism – offers a Bourdieusian analysis of the mediatized world of elite cosmopolitans, represented mainly by highly skilled expatriates of Scandinavian origin working for the United Nations in Geneva. It is shown how the autonomy of social agents is negotiated in relation to an increasingly mediatized communicational doxa, ultimately problematizing what it means to be privileged. The findings suggest that mediatization mainly constitutes an *indirect* cultural and material force. While mediatization substantially alters the social pre-conditions for accumulating "cosmopolitan capital" – that is, the resources for further social and geographical mobility – the appropriation and mastery of various media do not attain any symbolic value as such. The basic features of doxa thus remain intact, and are even reinforced through the more or less involuntary, or *osmotic*, appropriation and normalization of new media. From an epistemological view the chapter demonstrates how communicational doxa can function as a conceptual bridge between critical mediatization research and theories of social power in a mobile society.

Chapter 6 – Mediatization and Post-tourism – investigates urban exploration as a potentially counter-hegemonic structure of feeling. Urban explorers set out to discover, appropriate and document abandoned sites of industrial society (e.g., closed-down factories, hospitals and office buildings) and other places at the hidden fringes of the urban landscape. In their aim to achieve authentic and existentially charged experiences of sites that are (at least not yet) of any interest to mainstream tourism or heritage institutions they can be understood as *post-tourists*. This also means that urban exploration unfolds as a culturally distinctive practice, clearly marked by the mobile ethos of middle-class groups. The chapter scrutinizes the role of media during different phases of urban exploration and draws a picture of an inherently media-dependent movement that at the same time maintains a critical stance towards the spatial consequences of mediatization. This dialectical tension is detected in the overarching communicational doxa of urban exploration as well as in particular moments of *reflexive hesitation* that imbue photographic practices and various online sharing practices.

Chapter 7 – Mediatization and Gentrification – brings together some of the main threads from Chapters 5 and 6 by looking at the ambiguous articulations of mediatization within urban and rural gentrification processes. Gentrification is understood here as a structural and spatial expression of organized self-realization among the middle classes, which at the individual, biographical level is tied to the privilege of *elective belonging* (Savage, 2010). Although mediated connectivity and expressivity can be seen as integral parts of, sometimes even pre-conditions for, creative place-making practices in formerly less exploited urban and rural areas, they may also foster contradictory experiences of social pressure, frustration and stress. As the analyses reveal, these experiences may look different depending upon conditions related to habitus as well as urban/rural cultural-material structures. What unites them, however, is that the ability to master connectivity, to stay in control of mobilities and flows, stands out as a sign of privilege and a defining element of elective belonging.

Chapter 8 – Rethinking Mediatization, Mobility and Social Power – synthesizes the main arguments of the book and assesses their broader theoretical implications in relation to social and cultural theory. Particular attention is paid to the complex and so far under-theorized relationships between mediatization, individualization and globalization. Based on the preceding analyses of middle-class mobilities (Chapters 5–7) it is concluded that the appropriation of new media among privileged mobile groups mainly reinforces globalization processes, meaning that connectivities, mobilities and processes of cultural inclusion are enhanced. However, these processes are not always and not in all respects experienced as liberating. On the contrary, growing media reliance often makes it difficult for individuals and groups to achieve a positive sense of autonomy. Against this background, the chapter discusses the question of *counter-mediatization* as a potential reaction to the dominant orders of mediatization. It also discusses how the dialectic of mediatization turns such abilities as *dis*connecting, staying in place and maintaining the coherence of the lifeworld into increasingly valuable social assets.

Note

1   In my previous work, beginning with my PhD dissertation (Jansson, 2001), I have applied mediatization to analyses of social and cultural transformations in relation to various realms of activity; for example, everyday lifeworlds (Jansson, 2003a, 2013a, 2015a), tourism (Jansson, 2002a, 2007a), consumption (2002b), surveillance (Jansson, 2012, 2015b) and urban/rural transformations (Jansson, 2003b, 2005, 2010a, 2013b; Jansson and Andersson, 2012).

PART I
A CULTURAL MATERIALIST PERSPECTIVE OF MEDIATIZATION

2
MEDIATIZATION IS ORDINARY

In his 1958 essay *Culture Is Ordinary*, Raymond Williams reflects on his life trajectory from a small town in Wales to Cambridge, England and an academic career.

> The making of a society is the finding of common meanings and directions, and its growth is an active debate and amendment under the pressures of experience, contact, and discovery, writing themselves into the land. […] The questions I ask about our culture are questions about our general and common purposes, yet also questions about deep personal meanings. Culture is ordinary, in every society and in every mind.
>
> *Williams, 1989: 4*

Williams projects his own experiences onto the ongoing debates around culture in the age of high industrialization. He contends that neither the Marxist account of commercial exploitation and the passivity of the so-called masses nor the Leavisian critique of the mechanical vulgarization of fine arts as well as traditional culture seem to fit with his own lived experience. Modern culture, the culture of common people, he argues, is not dying. On the contrary, culture is continuously expanding and being re-negotiated in everyday life as well as through artistic experiments, both traditional and creative. Ultimately, we should come to realize that *culture is ordinary*, in the sense that it is always moulded through real life and thus retains its material presence in society.

Williams never spoke about mediatization. One might wonder what he would have thought about this rather awkward term that has been debated mainly in German and Scandinavian language contexts over the last decades (adopting the terms *Mediatisierung* and *medialisering*, respectively). Still, in this chapter I will argue that Williams's perspective, which he labelled *cultural materialism* (see Williams, 1980), has much to contribute to the debates around mediatization. The value of

Williams's work lies in its insistence on seeing culture and materiality as inseparable forces in social transformation, a view that speaks directly to the understanding of mediatization as a historical meta-process whereby society is reshaped under the influence of a broad array of media developments (see, e.g., Krotz, 2007, 2014; Couldry and Hepp, 2013; Lundby, 2014a). Cultural materialism positions mediatization as a transformational force that is inseparable from, rather than external to, the textures of everyday life. It is a perspective that helps to address two recurring problems in mediatization theory. First, it helps us discern where mediatization "begins", which is essentially a response to the dilemma of identifying *what is not* mediatization in the realm of culture and everyday life (see, e.g., Deacon and Stanyer, 2014, 2015). Second, and more importantly, cultural materialism brings forth the critical potential of the mediatization concept, a potential that has been remarkably absent from the discussions on mediatization so far, especially within the social-constructivist field.

The aim of this chapter therefore is to demonstrate and discuss the value of cultural materialism as a framework for articulating mediatization as a perspective for *cultural critique*. I develop this argument mainly through a discussion of Williams's concepts of *cultural form* and *structure of feeling*, but also in relation to his writings on *hegemony* (see especially Williams, 1974, 1977). These discussions lead to a critical view of mediatization as a broad societal transformation in which continuous everyday adaptations to, and negotiations of, media as socially amalgamated cultural forms also implicate the modification and emergence of structures of feeling. Williams's ideas around structures of feeling are worth relating to mediatization because the term builds a bridge between social/hegemonic power structures, largely intertwined with the circulation of media technologies and symbolic contents, and everyday lived experiences – that is, how it *feels* to live with media and ultimately to be dependent on them. As mentioned in Chapter 1, the empirical analyses of this book focus specifically on *mobile lives* as a mediatized structure of feeling.

I then expand my exploration of the cultural materialist approach to include Pierre Bourdieu's theory of *social fields* (see, e.g., Bourdieu, 1972/1977). Bourdieu's work is not usually termed cultural materialist. However, as Milner (1994:64) points out, if we conceive of cultural materialism as a method that regards cultural practices as material production and sees material structures as culturally meaningful, then there is good reason to consider Bourdieu's work as cultural materialist. Above all, he shares with Williams a strong focus on *culture as ordinary*, and power structures as reproduced through everyday *practical sense*. In relation to mediatization, I also investigate some interesting conceptual differences between Williams and Bourdieu that make their theories complementary. In particular, Bourdieu's notion of *doxa* (Bourdieu, 1972/1977, 1997/2000) provides a comprehensive platform for making sense of how the naturalization of various media technologies and practices (as cultural forms) is conditioned by the rules and resources of a social field, while at the same time posing a potential threat to these rules and resources. In order to establish the link between mediatization and Bourdieu's field theory I elaborate the concept of *communicational doxa*, which refers to the taken-for-granted communicative conventions and demands (including media practices) that regulate what it takes to be(come) a member of a certain field.

At the end of the chapter I engage in a synthesized discussion of how Bourdieu's perspective can be brought into dialogue with Williams's cultural materialism in order to advance mediatization research as a platform for cultural critique. The first step in this theoretical excursion, however, is to pinpoint why this kind of elaboration of research is needed at all. In the following section, therefore, I give a brief account of what I think is lacking in mediatization theory in spite of the rapid growth of mediatization research since the mid-2000s.

Where is cultural critique in mediatization research?

There are in social and cultural theory a few classic examples of how mediatization has been applied as a concept to identify a growing dependence on, and submission to, media. These examples, however, are for the most part located outside the mediatization debate as such. In the 1970s Jean Baudrillard used the term in his theory of the simulacrum, where mediatization meant "what is reinterpreted by the sign form, articulated into models, and administered by the code (just as the commodity is not what is produced industrially, but what is mediatized by the exchange value system of abstraction)" (Baudrillard, 1972/1981: 175–76). Mediatization here refers to a meta-level of symbolic circulation in which mediated signifiers have lost their connection to external referents, and thus to any deeper communication, and constitute rather a self-referential and culturally absorbing hyper-reality (see also Bolin, 2014). A similar example is found in Paul Virilio's (1995) pessimistic account of how "motorized" means of communication historically, and especially referring to the late twentieth-century American mass media complex, have come to govern people's worldviews and ultimately the world they live in. Mediatization is thus seen as a form of cultural dominance.

These accounts are, however, strikingly detached from the contexts of everyday cultural meaning production. Their mode of critique is primarily transcendental rather than immanent, and, while interesting as such, does not take into consideration the interpretative and creative capacities of "ordinary" people. Furthermore, they give us little space to make sense of the ambiguous feelings that often characterize the lived experiences of media culture. Interesting analyses of such ambiguities have been presented in recent years by a number of media researchers. Gregg (2008, 2011), for instance, shows how the normalization of "flexible work" among white-collar professionals, enabled by networked media technology, also leads to stress because of the dissolving boundaries between work, family life and leisure. Similarly, Turkle (2011) argues in *Alone Together* that our private media devices and the online realm of intensified mediated connectivity have created a shield that discourages us from confronting other people face-to-face to deal with complicated social and emotional matters. Allmer, in his book *Critical Theory and Social Media*, advances a Marxist perspective of the dialectics of social media based on surveys of young media users. One of his main points is that corporate social media maintain a use value based on people's extended abilities to interact and stay connected, but should at the same time be understood as social factories that have extended the logics of the factory to the Internet and "subsumed society and social activities into the capitalist process of production" (2015: 176). Accordingly, the young people in Allmer's study also feel that advertising and other techno-commercial imperatives somewhat constrain their sense of social autonomy (see also Allmer et al., 2014). These examples, and a number of others, testify to the contradictory, even dialectical nature of mediatization. But they do not refer to mediatization explicitly.

So where is cultural critique in the ongoing mediatization debate? The picture has two sides. One side is painted by Stig Hjarvard (2008: 113), who in his article "The mediatization of society" argues that by mediatization we should "understand the process whereby society to an increasing degree is submitted to, or becomes dependent on, the media and their logic". This is indeed a good starting point for unveiling the tensions between the emancipatory and restrictive forces of mediatization. Hjarvard (ibid.: 114) contends that while mediatization should be treated as a non-normative concept, "there is a general tendency in both research and public discussion to presume that institutions' dependence on the media is essentially questionable". Here he refers explicitly to the work by Mazzoleni and Schulz (1999) on the mediatization of politics as well as to the broader discourse of Habermasian theories of the public sphere in which the over-expansion of media power is associated with the disintegration of civil society. He further asserts that the issue of whether mediatization is "positive" or "negative" remains an empirical question that needs to be addressed in relation to specific contexts "where the influence of specific media over certain institutions is gauged" (ibid.).

Hjarvard's discussion identifies the presence, as well as the problems, of critical perspectives within mediatization theory. At the same time, however, it implicitly testifies to the lack of critical perspectives *pertaining* to the realm of culture and everyday life. The discussion revolves around transformations at the level of *institutions*, here defined in Giddensian terms, meaning socially demarcated areas of activity defined by a certain set of (formal and/or informal) rules and recognized resources (see Giddens, 1984). This is the other side of the picture: critical approaches to mediatization seem to be closer at hand for those working within the so-called institutionalist strand of mediatization research (see Lundby, 2014a). If we look at the collection of essays *Mediatization of Communication* (Lundby, 2014b), containing 31 chapters, we find that only one chapter explicitly mentions "power" in its title. The chapter in question is Asp's on the mediatization of politics, which among five key elements recognizes "adaptation as a process of social learning to a changing media environment" and "the media as constraints on actions" (2014: 349). These elements echo Hjarvard's (2008) view of mediatization as increasing institutional submission and dependence vis-à-vis the media. But we do not learn much about what is to live in a mediatized culture and society.

The main alternative to the institutionalist approach to mediatization is *social constructivism* (Couldry and Hepp, 2013), or what others have called the "cultural perspective" (Lundby, 2014a) or "media as world" (Bolin, 2014). This perspective has been advanced especially through research programmes conducted at the University of Bremen, led by Andreas Hepp and Friedrich Krotz. Hepp advocates that we should think of mediatization as the expansion of "media cultures", understood as "those cultures whose primary resources are mediated by technological means of communication, and in this process are 'moulded' in various ways" (2013: 5). In Krotz's (2007, 2014) work mediatization is explicitly outlined as a meta-process through which everyday practices of all kinds become intertwined with media. This is not a recent transformation, he writes, but "a historical long-term process that has occurred since the beginning of human communication" (2014: 132–33). As a consequence of mediatization, various areas of social life and communication are fundamentally altered. This, however, should not be understood simply as a technologically driven process. Rather, as Krotz argues, mediatization research needs to analyse the interplay between media change and social and cultural change, to see the human being as "a socially and communicatively constructed subject in society" (ibid: 140). Referring to Raymond Williams, Krotz also holds that this complex view of mediatization rests on the understanding of a medium as *both a technology and a cultural form* (ibid.: 144).

This approach is clearly aligned with what I suggest here, not least in the sense that it is concerned with the multi-faceted significance of media in everyday life, focusing not on single media or media logics, but on "extremely complex arrangements of different forms of media-based, communicative action" (Hepp, 2013: 17, italics removed). However, there are also problematic aspects associated with the social-constructivist view. The main problem is that nobody has yet tried to develop a framework for a cultural critique of mediatized everyday life. Krotz acknowledges this when he mentions (2014: 158) that mediatization research needs to have an "integrative and critical branch", addressing questions of privacy, exploitation, alienation and so forth, and taking into account the interplay between mediatization and other meta-processes (notably individualization and commercialization). The starting point for critical mediatization research, according to Krotz, should be that all mediated communication (in contrast to face-to-face-communication) involves a third actor who may exercise various types of power. Similarly, Hepp (2013: 143-4), in the final two pages of his *Cultures of Mediatization*, delineates three principles for advancing a critique of cultures of mediatization. Critical research should (1) focus on "the constructive process of cultural articulation" (notably the fact that "the media" are regularly constructed as "central" to culture and society); (2) focus on "the relation of cultural patterns to questions of power"; and (3) follow a multi-perspectival and comparative approach in which the articulations of different media cultures are taken into account.

However, none of the critical principles of Hepp and Krotz have been (to my knowledge) further developed or implemented, which means that there is still a need for analytical concepts that can make up a critical framework. We need tools for making sense of and articulating the composite everyday cultural and material power of mediatization without succumbing to overly static notions of institutionalized media logics (which do not fit the realm of everyday life) or analytically open-ended notions of everyday media saturation or "media life" (Deuze, 2011). The power of mediatization unfolds through very concrete processes of distribution (in terms of both material and cultural resources) as well as through uncomfortable experiences and feelings of dependence and restricted autonomy. We therefore need conceptual tools that can link everyday lived experiences to socio-material structures and locate empirical findings of everyday media adaptation and normalization within the broader historical panorama of mediatization. What we need, then, I argue, are *cultural materialist theories of the middle range*. In the following two sections I introduce Williams and Bourdieu as key representatives of such theories.

Williams: Mediatization through ordinary culture

Cultural critique necessarily begins with culture, and, as Williams says, *culture is ordinary*. Culture may look quite different in different environments and during different periods, but it always has something to do with the common meanings that flourish among people and guide them through their everyday lives. Culture is located in people's minds and bodies as well as in the structural arrangements and institutions that keep societies together. Similarly, when we speak about mediatization we are not referring to single media events or isolated experiences of certain technologies or texts. What we mean are the continuous changes and modulations of behaviours, worldviews and social relations within ordinary life that are, in turn, related to the structural normalization of certain media technologies and institutions in society. The relationship between mediatization and cultural change can therefore be described as *changes within the regimes of the ordinary*.

How can this help us in delineating mediatization as a form of cultural critique? Williams's poetic style of writing, and his concern with the complexity of lived experience, might seem at odds with any ambition to establish criteria for the identification and evaluation of media-related cultural change. However, his cultural materialism can be positioned as the general foundation on which to build an increasingly elaborated theoretical framework. In this section I argue, firstly, that Williams's basic notion of culture as ordinary provides a starting point for defining the negotiated nature of mediatization, and thus also for précising *what is not* mediatization. The key term here is *cultural form*, as elaborated in Williams's (1974) writings on television as technology and cultural form. As a second step I present Williams's (1977) *structure of feeling* as the intermediary social realm through which we can detect the broader cultural contradictions and ambiguities of mediatization; that is, the dialectical relationship between emancipatory and constricting forces. I also consider Williams's theory of *the hegemonic*, which leads us back to ordinary culture as a site not only of shared meanings but also of socio-political struggle and dominance. From this we may develop a critical view of mediatization.

In his 1974 *Television: Technology and Cultural Form* Williams tells two stories about the social normalization of television in modern society. One is the social history of television as technology (focusing on the actual invention and (social) engineering of the medium); the other is the social history of the uses of television. The point in telling these complementary stories is to show that the social impacts of new technologies are determined neither by scientific, political or industrial forces, nor by technology itself. Rather, we must understand a phenomenon like television as a continuously negotiated cultural form – a form that actualizes the inherent capabilities of technology while at the same time responding to social, cultural and economic pre-conditions in society. Williams's famous proclamation of popular broadcasting as a social product of a broader tendency in modern society, what he calls *mobile privatization*, is a vivid example of this dualistic view. The cultural formation and pervasive impact of radio and television cannot be understood without taking into account the fact that these technologies were introduced at a time in history when there was also a rapidly increasing level of everyday (auto) mobility (especially commuting) and general improvements in suburban family housing. Along with cars and electrified kitchen equipment, private broadcasting technologies were normalized as essential elements of mobile, yet home-centred middle-class lifestyles (and were soon also spreading to broader strata of society) (see also Spigel, 1992).

Williams's social history of television can be read as a tale of mediatization, projected through the lens of one particular medium. It is a tale that describes how the extraordinary impact of television occurred through the materialization of a particular cultural form; that is, domestic viewing practices corresponding to the modern need for rapid information flows *and* new forms of social communion. If television had not taken on a broader social significance and had not developed into a generally recognized cultural form, there would not be any reason to discuss this in terms of mediatization. As Marvin (1988) demonstrates, modern history is full of failed media technologies, inventions that have not caused more than a short-lived buzz or fascination among the general public. It is only when media technologies (individually and as ensembles) take the leap from the hypothesized or experimental realm, largely residing in the minds of engineers, politicians and business people, to the stages of lived experience that they become relevant to the history of mediatization.

This does not mean that mediatization only includes globally institutionalized media such as television. Nor does it mean that television constitutes a single or fixed cultural form. Clearly television is not the same thing today as it was in the 1970s, and Williams himself discusses the divergent cultural formations of television that emerged in British and American contexts. My point is that mediatization is a relevant labelling of social change only in as much as media are taken up as parts of ordinary culture; that is, when there have emerged more or less ritualized patterns of media use as well as shared understandings of particular media as indispensable – in short, when there is a cultural form. This may pertain to society at large or to particular social fields or (sub-)groups in society. When we speak, for instance, about the mediatization of academia, we may identify complex textures of media technologies and applications (e-mail, laptops, digital projectors, professional networking sites, and so forth) that together constitute a deeply felt order of practical necessity and social expectation. Such socio-technological orders testify to the interplay between technological capabilities and socio-cultural pre-conditions through which various media become part of the ordinariness of (academic) culture and are thus rendered a certain cultural form, which may in turn contribute to the alteration of more general orders of social and communicative recognition (cf. Figure 1.1).

Williams's concept of cultural form, then, can help us to explicate the situated dynamics of various sub-processes of mediatization (see Krotz, 2014). The concept reflects his broader framework of cultural materialism, which he originally defined as "a theory of culture as a (social and material) productive process and of specific practices, of 'arts', as social uses of material means of production (from language as material 'practical consciousness' to the specific technologies of writing and forms of writing, through to mechanical and electronic communications systems)" (Williams, 1980: 243). This formulation captures how he attempted to move beyond the Marxist over-determination of material relations and to see culture rather as an integral part of social power relations. Thinking of mediatization from a cultural materialist perspective means that we should consider the ways in which everyday media practices oscillate between creativity and productive processes, on the one hand, and growing dependence and structural reproduction on the other. This tension is an integral part of every cultural form, but articulated and experienced in different ways in different social settings.

This leads to the critical potential of cultural materialism. In *Marxism and Literature* Williams applies the Gramscian concept of *hegemony* to develop a view of culture that insists on "relating 'the whole social process' to specific distributions of power and influence" (1977: 108). Hegemony, Williams argues, moves beyond dualist theories of "reflection" and "mediation" as well as structuralist theories of "correspondence" and "homology", which tend to see culture as separate from the socio-material conditions of society. Hegemony captures the *internalization* of power relations in everyday culture and lived experience and is thus particularly well suited for making sense of dominance and subordination in advanced modern societies where such relations cannot be explained through simplified notions of, for instance, "manipulation":

> It [hegemony] is a whole body of practices and expectations, over the whole of living: our senses and assignments of energy, our shaping perceptions of ourselves and our world. It is a lived system of meanings and values – constitutive and constituting – which as they are experienced as practices appear as reciprocally confirming. It thus constitutes a sense of reality for most people in the society, a sense of absolute because experienced reality beyond which it is very difficult for most members of the society to move, in most areas of their lives.
>
> *Williams, 1977: 110*

The hegemonic thus positions ordinary culture not only as a source of communion (tradition) and innovation (arts) but also as a *structure of ordinary restraints* that people experience, but hardly reflect upon, in their everyday lives. Similarly, if we think about twentieth-century mediatization in the context of mobile privatization, as outlined by Williams, we see that it was a social transformation that involved the normalization of television as a cultural form (amongst other changes), which in turn established new spaces of everyday dominance and subordination. In general, people not only normalized these new perspectives of the world, confirming the cultural centrality of television (Couldry, 2003a), they gradually reorganized their whole way of life in relation to television and other social and infrastructural developments. The cultural and material boundaries of the ordinary were thus modified, encouraging expanded mobilities while at the same time invoking new forms of restraint.

In short, cultural materialism invites us to think about, and analyse, mediatization as a hegemonic transformation of culture and society. But it is not a one-way track of homogenization and intensified domination. As Williams (1977: 113) argues, "the most interesting and difficult part of any cultural analysis, in complex societies, is that which seeks to grasp the hegemonic in its active and formative but also its transformational processes". Hegemony harbours tensions, and so does mediatization. While dominant expressions of culture tend to be fixed and available for analysis, for example in literature, education and media, such patterns in some sense always speak of the past. They are the sedimented outcomes of creative energies emanating from a previous state. Practical consciousness, in contrast, is located in the present and is thus not fully graspable or available to classification according to simple categories. It refers to "what is actually being lived, and not only what it is thought is being lived" (ibid.: 112-113). This means that counter-hegemonic *structures of feeling* may evolve in between dominant cultural formations and lived practice. I propose later in this book that elite cosmopolitanism, post-tourism and gentrification constitute interesting sites for exploring such tensions.

Structure of feeling is a difficult term. As Milner (1994: 55) notes, its meaning was gradually altered in Williams's writings. From originally referring to "the culture" of a period, the generationally specific aspects of both art and lived experience (Williams, 1961/1965), it was later used to describe the emergence of counter-hegemonic movements in society (referring particularly to art and literature). Structures of feeling in the latter sense, and in the plural, refer to a "cultural hypothesis" (Williams, 1977: 132), or to the pre-emergence of cultural elements that anticipate changes in the more general culture of a society. Often such pre-emergences take their energy from everyday feelings of ambiguity and/or alienation in relation to the dominant culture, but they may also be initiated or reinforced by new opportunities for cultural expression and recognition. This is why structures of feeling become relevant to our understanding of mediatization. *How does it feel to live with mediatization?* What senses of frustration, stress, alienation, emptiness and dependence does mediatization give rise to? And what new opportunities for social and cultural transformation emerge from within mediatization itself?

In cultural materialist analyses of mediatization, I argue, we should apply structure of feeling on two different levels, reflecting the gradual re-interpretation of the term that can be found in Williams's work. First, we should think of structure of feeling as a cultural diagnosis of generally shared sentiments and values that may emerge around media (but not necessarily because of them) when these are normalized as cultural forms. We may argue, for instance, that mobile privatization brought with it new modes of togetherness and recognition related to mediatized suburban lifestyles that also reconfigured the general structure of feeling. This did not look exactly the same in all groups, but contained certain denominators that set it apart from what earlier generations had ever experienced. We can detect a similar shift today. The empirical identification of "media generations" (such as "generation analogue" vs. "digital natives") can be seen as an articulation of shifting modes of practical consciousness (Bolin, 2017). For instance, different generations tend to make different moral judgements when it comes to "appropriate" and "inappropriate" uses of social media and mobile devices (Bengtsson, 2015a). A sign of the social force of mediatization, accordingly, is that certain groups no longer feel "at ease" or "at home" with the media.

Second, we should refer to structures of feeling in their counter-hegemonic form, following Williams's later writings. We may then identify the emergence of alternative uses of media technologies as well as new cultural expressions that are reactions to the growing indispensability of media at large. These pre-emergences, whose cultural significance is always difficult to predict, arise as *reactions* to mediatization in its hegemonic shape (often intertwined with other meta-processes such as individualization, commercialization and globalization). A good example is the transformation of music listening during the twenty-first century. The recent achievements of Spotify can be traced back to a series of pre-emergences (including early file-sharing sites like Napster) that have their roots ultimately in changes at the level of practical consciousness. The status of counter-hegemonic forces is not always easy to pin down, however. In a situation where media technologies allow for a broader range of creativity and user involvement and where creativity is often an integral part of business models, as in dominant social media industries (e.g., van Dijck, 2012, 2013; Fuchs, 2014), the character of mediatization becomes increasingly multi-faceted.

This is not the place to pursue any longer excursions into emerging cultural movements or to discuss the more precise consequences of contemporary mediatization processes. My point is that the cultural materialist perspective can help us articulate mediatization as something ordinary *as well as* a site of negotiation and struggle both at the mundane level of everyday practice and within culturally innovative movements. Whilst Williams's conceptual framework is difficult to translate into measurable categories of mediatization it is a good starting point for developing a more comprehensive agenda for critical mediatization research. The focus should then be on how the cultural-material tensions between new creative energies (autonomy) and structures of ordinary restraints (dependence) unfold during different historical periods and in different social contexts, and how mediatization is thus related to prevailing power structures in society.

In adressing these questions, I will now turn to a related and more social science oriented approach: Bourdieu's theory of social fields. The main attraction of

Bourdieu's field theory in the context of mediatization is that it introduces context-sensitive tools for analysing the implications of mediatization in different areas of society. It thus adds an important horizontal power dimension to the basic cultural materialist approach that I have just outlined.

Bourdieu: Mediatization through communicational doxa

There have been relatively few attempts to elaborate Bourdieusian theory for the purpose of understanding mediatization, and even fewer arguments as to the validity of mediatization as a concept that could add value to the Bourdieusian framework. The most interesting attempt so far is Couldry's (2003b, 2012) analysis of the media as a "meta-field" that occupies a position similar to that of the State as described by Bourdieu (e.g., Bourdieu and Wacquant, 1992: 110–15; Bourdieu, 1996). From such a view the operations of media institutions and their agents should not be analysed primarily in terms of a distinct logic and form of capital ("media capital") comparable to those defining specialized fields like art, literature and other academic disciplines. Rather, media should be seen as conveyors of meta-capital that cuts across and contributes to the legitimation of more specialized fields. Couldry suggests that

> the media's meta-capital over specific fields might operate in two distinct ways: first, as Bourdieu explicitly suggests for the state, by influencing what counts as capital in each field; and second, through the media's legitimation of influential representations of, and categories for understanding, the social world that, *because of their generality*, are available to be taken up in the specific conflicts in *any* particular field.
>
> *2003b: 668, italics in original*

As a case in point, Couldry discusses the growing importance of media exposure and image creation that permeate a number of social fields today.

More recently, and logically I think, this line of thinking has led Couldry (2014) to locate the notion of media meta-capital within a theory of mediatization. As he argues, there are good reasons to conceive of mediatization as a meta-process that operates in *non-linear* and *transversal* ways, meaning that it exercises different kinds of influence within different fields depending on how media meta-capital affects the circulation and legitimation of specific forms of capital. This perspective grants us a sensitizing approach that moves away from more reductive understandings of "media logics" that tend to fall short when it comes to making sense of the manifold and fluctuating appearances of mediatization in different parts of social space. It also identifies a way of detecting how mediatization (re)produces power structures and thus helps us formulating a Bourdieusian cultural critique.

Still, Couldry's perspective shares some of the problems of the institutionalist approach. His focus is on media-as-institutions and their influence in terms of symbolic power. The same can be said about other recent efforts to integrate Bourdieusian theory in mediatization research, such as Rawolle and Lingard's

(2010, 2014) analyses of the mediatization of educational policies. The main concern is to chart the media's privileged position in shaping dominant discourses of the world and, by extension, the logics of fields and worldviews of ordinary citizens (see also Couldry, 2003a). In comparison to the State, however, media, and thus mediatization, entail much more than such symbolic-institutional processes as world description, prescription and legitimation. As we saw above, media attain a material appearance in the lives of social agents in the shape of continuously evolving technologies-as-properties that form the basis of, and amalgamate with, different kinds of social and cultural agency. The notion of "media meta-capital" is thus only valid within confined areas of what we may refer to as "the media"; namely; those institutionalized areas that show some resemblance to the State.

Without disregarding in any way the value of earlier institutionalist attempts to link Bourdieu and mediatization theory, what I want to introduce here is a more holistic and practice-based perspective that relates mediatization to the structural dynamics of everyday life as charted, for instance, in Bourdieu's (1979/1984) work on cultural taste and distinction. This means that mediatization is discussed in terms of the normalized, and growing, indispensability of media as cultural forms within the internal logics of social fields as well as their expansion into associated realms of social life. Linking this to Williams's cultural materialism, we may critically assess how appropriations of media within the *doxa* of particular social fields are conditioning, as well as conditioned by, shifts within broader structures of feeling.

The concept of doxa can be traced back to Husserlian phenomenological theory and its understanding of the lifeworld as an intersubjective realm of taken-for-grantedness (see, e.g., Schütz, 1962; Schütz and Luckmann, 1973). Doxa is the shared principles and norms of practice that keep communities together, making their members act in predictable ways that reproduce the order of the lifeworld. Doxa is thus a source of social security, granting a sense of belonging and placement so long as the individual adheres to the established order. Accordingly, doxa imposes restrictions on the autonomy of social agents through consent rather than direct force, mediating "the dialectic of objective changes and the agents' aspirations, out of which arises a *sense of limits*, commonly called the *sense of reality*" (Bourdieu, 1972/1977: 164). Bourdieu's view of doxa resembles Williams's description of hegemonic cultural relations. It thus leads us to problematize the ways in which social power-relations are maintained and evolve as normalized and broadly accepted orders of things.

> We need thoroughly to sociologize the phenomenological analysis of doxa as an uncontested acceptance of the daily lifeworld [...]. [W]hen it realizes itself in certain social positions, among the dominated in particular, it represents the most radical form of acceptance of the world, the most absolute form of conservatism. This relation of prereflexive acceptance of the world grounded in a fundamental belief in the immediacy of the structures of the *Lebenswelt* represents the ultimate form of conservatism.
>
> *Bourdieu and Wacquant, 1992: 73–4*

To act and relate to the world in "the doxic mode" is a matter of submitting to the rules, relationships and classificatory structures that constitute the social world of which one is already part. It means accepting that one's autonomy as a social agent is based on the recognition of one's propensity to act in line with doxa and thus to reproduce that same order that is a pre-condition for one's status as a social agent. We can see this mechanism played out in everyday life in the social sanctions that strike those who transcend the boundaries of what is acceptable or in the unease felt by those who enter social arenas where they do not belong. However, it becomes even more relevant when analysing the workings of social fields, whose nature is in a more clear-cut sense defined by "the rules of the game" and the circulation of specific forms of capital. As Bourdieu points out in a discussion of the artistic field, the autonomy of individual agents can only be granted so long as these agents act within the doxic confines of the field and thus submit to the preservation of its "purity" and difference in relation to other fields.

> Thus we discover that the autonomy acquired by artists, originally dependent for both the content and the form of their work, implied a submission to necessity: artists had made a virtue out of necessity by arrogating to themselves the absolute mastery of the form, but at the cost of no less absolute renunciation of function. As soon as they want to fulfil a function other than that assigned to them by the field, i.e., the function which consists in exercising no social function ("art for art's sake"), they rediscover the limits of their autonomy.
>
> *Bourdieu and Wacquant, 1992: 110*

The autonomy granted by a field is thus conditional and is ultimately an illusion established by the above-mentioned dialectic between objective reality and aspirations. Autonomy only exists when agents believe in the ordered arbitrariness of doxa and continue to make investments in the field through conformist practice. Bourdieu's special term for this type of belief is *illusio*, which is to be seen not as an order of reflexivity, he argues, "but of action, routine, things that are done, and that are done because they are things that one does and that have always been done that way" (Bourdieu, 1997/2000: 102). *Illusio* is the embodied sense of doing the right thing and being in the right place, thus grounding a tacit adherence to doxa.

I will now return to the question of media and communication. In the context of field maintenance, communication practices can be seen basically as another instance of "things that one does", but which cannot be done in just any fashion without jeopardizing one's membership of a community or field. I thus suggest that we think of *communicational doxa* as a sub-category of doxa that prescribes the ways in which social agents should communicate with one another, within and across fields, and with what means – that is, through what media. There are in Bourdieu's work many examples of the importance of communicative manners and how they are unconsciously adjusted to the requirements of doxa. He mentions the correction of accents when speaking to persons of higher rank and the choosing of appropriate language in multi-lingual situations (e.g., Bourdieu, 1997/2000: 184;

see also Goffman, 1959). Furthermore, in *Distinction* Bourdieu (1979/1984) unveils an entire universe of unspoken rules that govern which means of communication different class fractions prefer, or find necessary; how different social groups communicate about everyday matters like cooking and bodily exercise; and how exchanges between agents within a specific field are accompanied by particular codes of communicative conduct. The fact that communication is part of doxa means that it attains classificatory power in a dual sense; on the one hand, through the classification of various goods and practices that are associated with the field (normally in the doxic, conservative mode), and, on the other, through the self-classifying recognition of, and submission to, doxa itself.

By drawing on these preliminary statements on communicational doxa it is possible to bring forth a Bourdieusian understanding of mediatization. If we define mediatization as a meta-process that is realized through the taken-for-granted indispensability of, and adaption to, technologies and institutions of mediation, the connection to communicational doxa is not far-fetched. When media become integrated in doxa it means precisely that they enter the realm of taken-for-granted order and necessity. The ways in which agents relate to hands-on technological features as well as institutionalized media logics then seem natural and attuned to the general expectations of doxa. A key advantage of conceiving of mediatization in this way is that the appropriation of "media" is seen as interwoven with, and inseparable from, social and cultural processes at large. Media are woven into the prescribed ways of doing things, which means that their meanings are also moulded through doxa. In other words, returning again to Williams, communicational doxa is directly related to the continuous shaping of media as cultural forms.

Speaking of "media dependence", then, does not refer to a type of dependence that is in any way "natural" or objectively linked to media as such; rather, it *seems* natural precisely because social agents put their belief in doxa – that is, *illusio* – in order to maintain a sense of autonomy. Doxa functions as a legitimation of media dependence under the auspices of granting further autonomy to agents who consent to the communicational doxa of the field.

What emerges here is a view of mediatization that is not restricted to the symbolic power of media institutions, but takes into account the materiality of media. It resonates not only with Williams's cultural materialism, but also brings to light what Silverstone (1994) calls the "double articulation" of media (see also Livingstone, 2007). I would even argue, in line with Hartmann (2006), that we can speak of a "triple articulation", meaning that media become part of communicational doxa in three different ways. First, they are integrated as *technics*. This means that they become indispensable in their capacity to establish certain relationships between the individual and the world, such as embodiment relations (extensions) and hermeneutic relations (representations) (see Ihde, 1990). Second, media may become part of doxa as *properties*; that is, as classified and classifying symbolic markers that are seen as required possessions for expressing the identity of an institution or agent. Third, the interweaving of media practices and doxa creates dependences through *textures*. This means that media become an integrated part of the taken-for-granted material environment and temporal rhythms of everyday life that normalize certain expectations regarding the positionality and regularity of media practices and things (Jansson, 2006, 2007c). As I will discuss further in Chapter 3, cultural form includes all three aspects.

The analytical potential of communicational doxa is strongest when implemented in relation to fields, because it can then be linked to a specific "game" where a certain type of capital is at stake. This is the primary meaning I want to ascribe to doxa. My suggested adaptation of Bourdieu may then shed light on one of Couldry's (2014) key concerns; namely, how mediatization potentially transforms the limits and logics of fields. The cultural materialist perspective suggests, however, that we study such changes, involving for example the overlapping of private and professional time-spaces, "from below"; that is, at the level of everyday practices and materialities rather than from above. This should not be taken as a dismissal of Couldry's points. Rather, it is a request (a) to conceive of media (and thus mediatization) as much more than just (the adaptation to) institutions of symbolic power, and (b) to engage in a Bourdieusian epistemic translation that more sufficiently accounts for the phenomenological traits of his theory (cf. Myles, 2004). It is only when we account for the mundane power of communicational doxa – that is, its *cultural ordinariness* – that we will be able to see how mediatization is socially realized and shaped through embodied practice.

Conclusion: Mediatization as cultural critique

In this chapter I have introduced a cultural materialist perspective of mediatization. I have argued that it is from such a perspective that we can sharpen the demarcations of mediatization as it unfolds in culture and everyday life. This in turn provides a platform for cultural critique. The critical component of mediatization ultimately boils down to the dialectical interplay between social autonomy and various forms of media-related dependence (see Chapter 1). As Krotz (2014) points out, the inclusion of media within communicative practices implies that there are third parties of various kinds influencing the process. However, the influence of such third parties does not substantiate a critical view of mediatization *per se*; it is only when media have attained a normalized presence in day-to-day life, when they have become more or less indispensable, that can we reasonably speak of dependence. This is where the cultural materialist approaches of Williams and Bourdieu become useful tools in critical mediatization research. They both account for how social power relations of different kinds are formed and negotiated in practical consciousness and thus have tended to evolve without much overt resistance. The meta-process of mediatization entails a compound of hegemonic (trans)formations through which media (in their different shapes) come to occupy culturally and materially taken-for-granted functions in ordinary life.

Cultural form is the key concept here. It is the main link by which Williams and Bourdieu can be connected to mediatization research. As discussed above, the social significance of mediatization resides in the establishing of media as cultural forms. This does not mean, however, that we should isolate mediatization processes to

Mediatization 1	**Mediatization 2**
Communicational doxa	Ordinary culture

Social field // Doxa ---- MEDIA AS CULTURAL FORM ---- Hegemony // Structure of feeling

| **Bourdieu** | **Williams** |

FIGURE 2.1 Cultural materialist interpretations of mediatization.

studies of individual media or that the aim of mediatization research is to establish clear-cut cultural forms. The point is rather to see how media in their overlapping cultural forms are culturally and materially integrated in different social contexts and how *through their indispensability* they become constitutive of social power relations (see Chapter 3).

In this regard Williams and Bourdieu give us two different, potentially complementary, directions for cultural critique (see Figure 2.1). In Williams, cultural form connects media to the (trans)formation of hegemonic relations in society, meaning that hegemony is at the same time an integral part of cultural form and is potentially challenged by media change (e.g., new technology) and/or the emergence of counter-hegemonic structures of feeling. In Bourdieu, cultural form can be associated with the concept of doxa, which refers to the taken-for-granted and embodied norms and values of a social field. Doxa thus holds a similar position to hegemony, but operates within a more confined area of social relations. It is at the productive intersections of cultural form vs. hegemony (Williams) and cultural form vs. doxa (Bourdieu) that we can detect the ambiguous force of mediatization. This is where the normalization of various media brings about not only expanded spaces of opportunity but also a sense of bounded reality, the acceptance of "things as they are" and "things one just does". In order to illuminate these conditions I have in this chapter advanced the intermediary terms *ordinary culture* and *doxa* – more specifically what I call *communicational doxa* – as the lenses through which we may look at mediatization from a cultural materialist perspective.

The theories of Williams and Bourdieu obviously lead to different but epistemologically related interpretations of mediatization. What can be gained from combining them in critical research? Is it at all possible? The strength of the Bourdieusian approach is that it provides an analytical framework for sorting out how mediatization unfolds in different fields; that is, how and why cultural forms are differently attuned depending on what forms of capital are at stake. There is no such apparatus in Williams's work; however, he integrates an important emotional component. Through structure(s) of feeling we are able to account for how the directly experienced restraints and pressures of mediatization may constitute elements of broader social transformations (the "culture" of a period or a general population) or give rise to explicitly anti-hegemonic movements. Bourdieu's perspective is comparably devoid of emotionally oriented analyses; a formulation like "feel for the game" (Bourdieu, 1980/1990: 66), for instance, points rather to socio-cultural pre-dispositions (*habitus*) than to the kinds of feelings modern culture (and mediatization) actually evoke.

We are thus dealing with cultural materialisms that have much in common but articulate slightly different aspects of society. The value in bringing them together in mediatization research emerges especially if we want to formulate a cultural critique that links the lived experiences and social trajectories of individuals and groups to the broader developments of culture and society. Such a critique should start out from the ordinary, since this is where mediatization finds its meaning.

3

WHY ARE MEDIA INDISPENSABLE?

Without my mobile I would feel like I was missing something. I would miss the contact with the Internet, yes the whole information society. I mean, if I'm in town without my phone, then I'll have to wait until I get home before I can check out what's happening in the world. So I would feel like being left behind, strange as it might sound, but I wouldn't be updated until I got home.

33-year-old man from a mid-size Swedish town

One of the clearest expressions of mediatization is the pervasiveness with which numerous media have become materially indispensable to people in their daily lives. Mediatization involves a variety of everyday material transformations, but a common denominator is the experience of *living with media things* as seemingly natural elements of the lifeworld (Schütz and Luckmann, 1973). People have literally (re-)arranged their life environments in response to the appearance (and disappearance) of media. As empirical studies have shown, this "media embeddedness", which in most cases today implies a life online, is also an emotional matter (e.g., Paasonen, 2015). The indispensability of media things, and thus the material force of mediatization, becomes particularly obvious on occasions of *dis*connection due to absent or malfunctioning media technology. For most of us, leaving home in the morning without the mobile phone would lead to feelings of frustration, even emptiness, a sense that something was lacking. Trying to log on to a wireless network without success may evoke similar feelings. Twenty years ago people were largely unaware of this particular type of experience, relying instead on other media and governed by other structures of feeling. Media come and go, and the face of mediatization changes.

Media things do not become indispensable by themselves, however. There are no media (if we think of "media" as various means of communication operating through certain symbolic codes) whose social success was a *fait accompli* at the time of their invention. Over the years, many technologies have failed to have any major social significance (Marvin, 1988; Gitelman and Pingree 2003), while others have rapidly fallen into obsolescence due to various contextual circumstances (cultural, economic, technological and political) (Löfgren, 2009; Acland, 2007). As we saw in Chapter 2, media (like other technologies) become indispensable only when practical affordances are brought into a meaningful relationship with pre-existing, or emerging, cultural-material conditions, thus giving shape to a particular *cultural form* (Williams, 1974). This is an important reminder of the socially oriented nature of mediatization research: we must never isolate the significance or impact of the media from surrounding processes in society.

Based on the cultural materialist perspective, as outlined so far, we can make three initial statements as to what mediatization *does* and *does not mean* in the context of material transformation. First, mediatization cannot be described merely as a linear process of *material accumulation* in which our everyday spaces become increasingly occupied by, or cluttered up with, *more* media technologies (from books and letters to television sets and smartphones). While many life spaces are indeed marked by a process of escalating media abundance, ongoing technological developments also point to the integration (or re-mediation) of a multitude of old media forms, as well as services, within digital *transmedia* devices and platforms. Books, newspapers, magazines and films can now be downloaded or streamed and do not have to occupy material space in the same way as before. The emergence of transmedia platforms and entire transmedia environments also brings with it the successive marginalization or elimination of many "stand-alone" media forms, such as CD-players, radio receivers and analogue cameras.

Second, mediatization does not refer only to people's celebration of, and longing for, new media stuff. As a meta-process, mediatization also includes social, cultural and ideological *resistance* to such consumption practices. Not everyone would like to have six television sets or five laptops at home, even if it were economically and spatially possible. Not everyone welcomes the increased reachability and monitoring enabled by mobile, interactive media. Let us consider, for example, the following interview extract that refers to altered, mediatized working conditions in a Swedish factory:

> In the new factory there are mounted cameras, about ten cameras, which overlook the whole production in the new hall. And those who built the new factory can watch it, in Slovenia. [...] Right at the steering unit there was a camera pointing straight down on us, and we never understood why it had to be there, so we poked it upwards a little, just so it couldn't see us. Because it felt like nobody trusted us.
>
> *26-year-old man from a small Swedish town*

One can of course debate whether such acts of resistance are to be considered integral parts of mediatization or as expressions of other, competing processes in society (cf. Deacon and Stanyer, 2014, 2015; Hepp et al., 2015). The difference can be hard to tell. I argue, however, that to the extent that certain forms of agency, cultural expression or counter-hegemonic movements – that is, structures of feeling (see Chapter 2) – emerge as direct reactions to the felt over-expansion of media, these should be taken as evidence of the meta-processual character of mediatization. (I will return to the question of *counter-mediatization* in Chapter 8.)

Third, the material and spatial dimensions of mediatization cannot be understood only, and perhaps not even most fully, through analyses of the actual *material presence* of various media. As mentioned in Chapter 1, one must also take into account how various places and practices are materially *adapted to* the existence of media. Home environments are not only filled with media, but also accommodated to particular forms of media consumption. In camera-monitored work places, such as the one just described, employees probably do not feel that such surveillance is necessary, but nonetheless have to adapt their practices to the potential telepresence of distant others.

Indispensability can thus be understood as *a bonding force between social agents, technologies and the world*. But *how* and *why* do "media things" become indispensable (and eventually obsolete)? What forms of dependence and adaptation do they give rise to? These are the questions we should ask. Ultimately, they demand complex answers that must take into account the contextual conditions in which media are actually put into use. The relationship between indispensability and adaptation may look very different depending on the type of technology and socio-cultural context. Such situated analyses are presented in Part II of this book. In this chapter, my ambition is more theoretical in nature. Building upon, and extending, the cultural materialist foundation introduced in Chapter 2, I propose to introduce a systematic approach to studying the social construction of material indispensability. In addition to the basic influences of Williams and Bourdieu my endeavour is guided by the works of Don Ihde and Henri Lefebvre.

What I suggest are three complementary levels of analysis, each offering a certain array of analytical entry-points for empirical study. At the first level I discuss media as *technics*, following Ihde's (1990) phenomenological view of technology and the lifeworld. At the second level, I look at media as *properties* in a Bourdieusian sense (Bourdieu, 1979/1984, 1997/2000), addressing the cultural shaping of indispensability. Finally, following Lefebvre's (1974/1991) understanding of *textures*, I discuss how media, as both technics and properties, become part of the felt cultural-material fabric of everyday life. Taken together, these analytical levels contribute to a holistic understanding of indispensability, helping us to unpack the various articulations of media as socially recognized cultural forms. The different articulations of cultural forms correspond to relatively stable (albeit contextually shaped) functions, classifications and time-space locations. They also correspond to different forms of dependence, as I point out in the conclusion to this chapter.

Throughout these discussions, and in the concluding part of the chapter, I pay particular attention to the ways in which the face of mediatization is currently changing owing to the shift from mass media technologies to transmedia technologies (see also Jansson, 2013a). As noted in Chapter 1, the categorical distinction between "mass media" and "transmedia" operates as shorthand for a bundle of digital developments (including, for example, convergence, interactivity, spreadability and miniaturization) that at the level of everyday life unfold as a gradual shift from stand-alone media fixtures to increasingly *integrated* and *flexible* polymedia environments. This transformation would suggest that indispensability is brought about under altered conditions, exposing partly new features. Accordingly, in this chapter I will position the analyses of this book in relation to the genealogy of media developments; my focus is precisely upon the ambiguous, or hybrid, orders of technology that emerge at the intersection of mass media and transmedia (see, e.g., Chadwick, 2013). While mediatization is not determined by technological shifts we certainly need to take into account how these shifts shape the pre-conditions for social and cultural change.

Media technics

My starting point for explicating how media become indispensable is Ihde's (post-) phenomenology of human–technology relations (1990: 72), and what he later developed into a "material hermeneutics" (Ihde, 1999). Ihde's work is concerned not just with how individual subjects (notably within the history of science) make sense of the material world, but also, and more importantly in the present context, how material objects as *technics* come between the world and the interpretative subject. His work is thus doubly material and helps us to bring the cultural materialist approach closer to actual technology. The general point of Ihde's theorization of technics is that these in various ways affect how people relate to and experience the world. His ideas are thus in keeping with those of medium theorists like McLuhan, even though they do not refer to media specifically. In comparison to medium theory Ihde's perspective also integrates a stronger focus on the hermeneutic and phenomenological implications of technological change, which provides for a more culturalist understanding of how different technologies shape (qualitatively) and enhance the mediatization meta-process.

In *Technology and the Lifeworld* Ihde (1990) introduces four principal sets of "I-technology-world" relations that will frame the following discussion: *embodiment relations*, *hermeneutic technics*, *alterity relations* and *background relations*. In embodiment relations, which are typically represented by optical technologies such as eyeglasses, the world is perceived through a technology whose presence is barely noticed or reflected upon by the subject. Eyeglasses, if they function well, "withdraw" from the wearer's experience of the world. Embodiment relations may thus be described as "the symbiosis of artifact and user within a human action" (Ihde 1990: 73). This means that the user and his/her tool or equipment become one, as in contexts of long-developed relations of handicraft (hammer, knife, etc.) or sports (skis, racing car, etc.). As Ihde points out, the dream of seamless body–technology relations has recurred throughout modern history, giving rise to utopian as well as dystopian prophecies of human cyborgs. However, media technologies have rarely managed to occupy such a symbiotic, invisible relationship with the body and the senses. Probably the telephone is the best example of a medium of communication whose technological functioning and material presence "withdraw" during the act of use.

The second set of human–technology relations is hermeneutic technics. Here we encounter the type of relations that have most generally marked mass media society. In contrast to embodiment relations they involve some act of reading in which technology is positioned as the interface through which the user can read the world. This is to say that hermeneutic technics, such as maps, charts and written texts, provide representations of space (cf. Lefebvre, 1974/1991), referring either to existing places or to fictional or planned spaces. In ideal cases, when hermeneutic technics are working smoothly, the user does not reflect on this interface even though the object of perception is precisely the technology as such rather than the world itself. One could say that a different type of symbiosis or transparency occurs, one between technology and the world, when the user enters the representational realm through the praxis of interpretation. As Ihde (1990: 82) explains, "textual transparency is hermeneutic transparency, not perceptual transparency".

This means that the technology, whether we speak of a thermometer, a newspaper or an industrial switchboard, becomes transparent and integrated into the taken-for-granted lifeworld only if the user possesses the appropriate hermeneutic skills; that is, masters one or several codes. The relationship also depends on the user's trust in the mediating capability of the technology. Transparency is immediately threatened if the reader does not find a certain scale trustworthy, for example, or suspects that information is incomplete or biased – a problem that has been scrutinized extensively and from different perspectives in media and communication studies ever since Shannon and Weaver (1949/1963) introduced their influential model of radio transmission.

The third and fourth variants of human–technology relations – alterity relations and background relations – are distinct from the previous ones in the sense that technology does not mediate between the individual and the world in any significant way. Nevertheless, these types of technologies are essential to the composition of modern lifeworlds. In the case of alterity relations, technology itself becomes the object of attention; the user is not given access to any world other than the imaginative space of technology itself. This is also an often mythologized or ideologically saturated (lived) space. Ihde mentions several different forms of alterity relations: the personalization of technology, through which artefacts are fetishized or sacralized; the othering of technology as something to master or contest; technology as a toy or object of fascination; and technology as something to interact with as a competitor. As Ihde argues, many technological innovations in history have attained the status of sacralized objects of fascination before being turned gradually into more mundane objects for daily use. This condition also holds true for many media technologies (Marvin, 1988; Mosco, 2004).

The final variant, background relations, is different from alterity relations in the sense that technology is not placed at the centre of attention, but operates in the background of other practices. In background relations, technologies function as "texturing" devices (Ihde 1990: 109) for creating certain environmental experiences (visually, audibly or materially), either within open spaces or as a way of generating spatial encapsulation (Jansson 2007a, 2007b). Background technologies thus attain the position of "an absent presence as a part of or a total field of immediate technology" (Ihde 1990: 111). This does not mean that they are neutral or less significant in the lifeworld than focal technologies, however. As Ihde argues, background technologies speak to different types of texturing affordances and often "exert more subtle indirect effects upon the way a world is experienced" (1990: 112). A case in point is the taken-for-granted but nonetheless crucial presence of audible background media in commercial spaces such as department stores.

There is an important common denominator in these four variants, pointing to the very core of the mediatization meta-process. Ihde's phenomenology of technics, which I have introduced just briefly here, clarifies in a systematic way how experiences of indispensability, and the necessity of adaptation, run parallel to the naturalization of media devices in the lifeworld. This does not mean that a particular technology in fact, or in any fundamental sense, becomes indispensable to social life or human existence just because its existence is taken for granted. However, the more a particular medium is taken for granted and the more it becomes transparent as technology, the more difficult it becomes to exclude it from the practices of day-to-day life. It is possible to assess, at least in a tentative manner, the significance of mediatization according to each of the four human–technology relations mentioned above and gain a more systematized understanding of how the indispensability of media evolves as a cultural-material phenomenon of our times. Even though Ihde's systematization preceded the vast expansion of mobile media technologies, it is particularly apt for clarifying how the introduction of networked, portable computers, touchpads and smartphones has propelled the mediatization meta-process into a sub-stage of transmediatization.

Transmedia technics

In a research project carried out in 2010–12 we asked our respondents which media technologies were the most important in their lives, and why (see Jansson, 2014). Most respondents (without taking this as general evidence) mentioned television, laptop and/or mobile phone. The latter two were chosen for their portability and functional versatility, thus highlighting the social significance of the technological leap from "ordinary" mobile telephones to smartphones (basically portable computers). The original transparency of telephone technology that Ihde talks about, the propensity of technology to "withdraw" through embodiment when talking to somebody, is combined with both portability and a number of other human–technology relations. The smartphone, and related platforms, thus represent technologies that cannot be categorized according to just one of Ihde's four variants, but involve processes of naturalization and disappearance at different levels, making them increasingly indispensible omnibus devices. As Wise argues, what is new about "the clickable world" is not disappearance as such, but "the scale of the disappearance, and the power the attenuating technologies potentially have over our lives" (Wise, 2012: 162).

Still, in order to systematize our discussion, we may go back to Ihde's four sets of "I-technology-world" relations and look at them separately. First, the fact that technological miniaturization makes it easier to carry, even wear, digital communication devices close to the body suggests that a whole new range of embodiment relations have emerged. Even though most functions embedded in, for example, smartphones require some kind of interaction via an interface, and thus imply hermeneutic relations, the experience of "nakedness" when not wearing one's mobile indicates that the very habit of having permanent, and instant, access to contacts, information, entertainment and so on via the online realm results in a sort of technological embodiment.

Second, the development of new software applications and refined interfaces has contributed to the transparency of hermeneutic relations and thus provided a sort of *lubricant* for mediatization processes. The installation of "smart" digital devices, ranging from smartphones to smart television sets, rarely requires any separate instruction manuals; users are guided through the installation process and can start using the new device within minutes. There is thus less hermeneutic work and fewer arbitrary learning processes involved in getting started. Furthermore, the iconography of smartphones and touchpads has a more direct code than the older, text-based menus and commands that dominated the digital realm until just a few years ago. Today, even one-year-old children quickly learn how to master these devices and navigate between different functions and contents.

In addition to this, the interactive nature of many applications, as well as online search engines and commercial websites, integrates algorithms that to some extent enable technologies to adapt to the user and his/her habits and preferences. As users we encounter offers and recommendations based on aggregate data that assign us to certain patterns and categories. This means that it becomes easier to find relevant information and services provided that we submit to a certain degree of surveillance (e.g., Andrejevic, 2007, 2014). Our view of the world, which the (mass) media have traditionally provided via news and other forms of content, is thus to a growing extent fused with a view of interactive "service spaces" (banking, e-commerce, etc.) and simulated views of ourselves, our habits and preferences. Transmediatization is intertwined with the emergence of an "algorithmic culture" (Striphas, 2015). We do not merely "*read* ourselves into any possible situation without being there", as Ihde (1990: 92, italics in original) puts it, but also track ourselves, and even start developing our tastes and lifestyles according to "nudging" applications (Wilson, 2012) and various forms of "curating by code" (Morris, 2015).

The altered shape of hermeneutic relations sustains indispensability in two interrelated ways: on the one hand, through adaptable software and simplified interfaces that make technology increasingly transparent, and on the other hand through the opening of a multitude of worlds via one single (and potentially interconnected) media device. While these factors do not in themselves explain (trans)mediatization, they are important for its lubrication.

Third, smartphones and associated devices blur the lines between hermeneutic technics and alterity relations. As already indicated, the types of "worlds" that these technics provide access to are increasingly diverse, and some of them are also more or less self-contained and self-referential. For instance, many lifestyle applications that encourage users to track and improve their performance, typically in sports, are designed to enable a significant degree of playability and spreadability (see Jenkins et al., 2013). This means that users enter into a world of play and competition, which on the one hand refers to a social world outside techno-space (and thus can be seen as a hermeneutic relation), but on the other hand contains modes of representation and attention-building that are more akin to alterity relations. Besides the fact that new technology may occupy an almost sacred position within the lifestyles of certain groups and individuals, related to novelty and brand value, an entire new world of game and play is created. These worlds can also be accessed almost anywhere and at any time thanks to the portability of new online devices, making these devices indispensable for "killing time" while waiting or travelling. These examples highlight the complexity of indispensability and clearly illustrate how this regime of mediatization is tied to both pre-mediations (conceived space) and the successive normalization of new social practices (lived space) (Lefebvre, 1974/1991).

Finally, media have a long history of generating and entering into background relations, in private as well as public spaces. Perhaps radio and other audible media have had the most prominent role in giving a certain "feel" to spaces and situations allocated to a range of social functions and practices (Tacchi, 1998). Here the main key to extended indispensability is whether such background relations involve a sustainable form of amalgamation between media use and the other functions, or not. As Schulz (2004) argues, amalgamation is one of the basic sub-processes of mediatization today, and it is not limited to relations through which media technologies produce socially shared environments. Again, the portability and versatility of new media devices enable single users to generate their own, technologically invisible soundscapes through which they can experience the world around them, for example while exercising (see Bull, 2001, 2007). This background relationship becomes a mode of being alone together with others (see also Turkle, 2011).

One can of course debate whether this generating of private, encapsulated spaces is a valid sign of material indispensability or not. Would it not be possible to dispense with media under such relatively exclusive conditions? Would it not be possible to exercise without listening to one's favourite music, for example? Questions like these are ultimately tied to moral and philosophical concerns and the dilemma of what constitutes an actual need among human beings – materially, socially, or in other ways. They can also be illuminated through considerations of everyday rituals. The amalgamation of private media technologies and other practices in the creation of background relations constitutes a good example of how certain individual activities are *ritually adapted* to the material existence and affordances of the media. I return to these issues in relation to media textures below.

Media properties

As demonstrated in the previous section, mobile transmedia technologies (compared to singular media) can be incorporated within the lifeworld in increasingly complex and open-ended ways. This should not be taken as a techno-deterministic view, however. Even though technologies are often significant in themselves, notably by means of their "disappearance", the actual magnitude of mediatization can never be estimated or understood without also taking into account the contexts, or social lifeworlds, within which particular "I-technology-world" relations materialize. In other words, "media things" are much more than technics. To a significant degree they are also *cultural properties* that may be appropriated or rejected on the basis of cultural values as much as on their value as functional assets. This is to say that our key concept, indispensability, is to be seen partly as a cultural construct whose phenomenological status fluctuates according to structural conditions.

Here Bourdieu's work on taste, doxa and practical knowledge consolidates the bridge between phenomenology and cultural materialism. A good example is his notion of "the taste for necessity" (Bourdieu, 1979/1984). In Bourdieu's analyses such an orientation is found primarily among the working classes, where the habits and preferences of social actors often remain stable even though their material conditions have altered. The inclination to demand considerably less than what is economically affordable implies that the taste for necessity is "operating out of phase, having survived the disappearance of the conditions that produced it" (ibid.: 374). In other parts of social space, however, the force of habitus – the invisible force of socially inherited dispositions – may look considerably different. Among mobile middle-class groups, particularly the "new bourgeoisie", one can discern conditions in which subjects have a vested interest in expressive consumption. This is partly due to the need to acquire "correct" lifestyle attributes that can match the standards of one's social aspirations (see Chapter 4). It is also, and at the same time, connected to a social desire for "ethical retooling" (ibid.: 310) of the economy as such. The interests of emerging middle-class fractions benefit from the continuous production of symbolic and social needs, a hedonistic morality based on consumption practices that reject the traditional ethic of sobriety, saving and accumulation.

If we combine these lines of thinking with Bourdieu's general argument regarding *economic* versus *cultural capital*, we can conclude that the social judgement of such phenomena as "necessary" and "indispensable" may fluctuate not only in terms of the *extent* to which individuals and groups are inclined to *appropriate* new media things – that is, make them their properties (Bourdieu, 1997/2000: 134) – but also the *types* of things they regard as desirable and/or necessary in their lives. Most of the time these judgements are not reflexively developed, but are integral to the lifeworld itself, structured by the force of habitus. Processes of appropriation are thus governed by practical knowledge and inform the structures of classification that provided the conditions for cultural judgements in the first place:

> [P]ractical knowledge is doubly informed by the world that it informs: it is constrained by the objective structure of the configuration of properties that the world presents to it; and it is also structured through the schemes, resulting from incorporation of the structures of the world, that it applies in selecting and constructing these objective properties.
>
> *Bourdieu, 1997/2000: 148*

This perspective adds a contextualizing layer to Ihde's phenomenological view of technics, stressing that the constitution of technological relationships partly depends on whether they can be legitimized within a certain socio-cultural order of recognition or not. From this it follows that while the economic epicentre of mediatization – that is, mediatization seen as a hegemonic and materially expansive process, as discussed in Chapter 2 – is located in those parts of social space where the production of new mediatized needs are deemed socio-culturally beneficial (typically within the mobile middle classes), there are also social sites where processes of extensive media appropriation are met with moral and cultural scepticism, and where the functionalities of certain new media technologies may collide with doxa and practical knowledge (see Jansson and Lindell, 2015).

As a case in point, let us consider the following quote, taken from an interview with a 35-year-old man with relatively little education living in a small Swedish town:

> We have six TV sets. Three upstairs, one in the living room, one in the base-ment and one downstairs in the playroom. Only two of them are connected to the cable, for watching television. One is for video and the home-theatre, one is for TV-gaming in the basement, and the kids have one each with DVD players in their rooms. And then one in our bedroom where we can watch television, just like in the living room.

It may seem quite extraordinary that there are six television sets in the one household. To what extent are they actually indispensable? Probably this family could keep up a fairly comfortable life with just one television. At the same time it is important to note that this man is able to rationalize these "I-technology-world" relation, as each television set performs a different function: gaming, videos, children's DVDs and TV programmes. Older machines are moved gradually to more peripheral places in the home and used for more limited purposes – and so, one might infer, become less and less indispensable (see Löfgren, 2009). But these are not value-free judgements. They can be articulated only to the extent that the social agent feels that his lifestyle is reasonably attuned to what is deemed socially legitimate within his circle. In a different social setting, for example one that obeyed a moral economy (Silverstone et al., 1992) marked by the possession of greater amounts of cultural capital, such a mode of legitimization would have been less likely. The cultural scepticism towards television in general, and excessive television watching in particular, which also manifests itself through the material shaping of households – such as the placement, size and quantity of television sets and associated appliances (video recorders, satellite dishes, etc.) – has been reported repeatedly in studies of broadcast audiences (see, e.g., Moores 1993, Jansson 2001; Bengtsson, 2011). Furthermore, it follows from the autonomizing logic of cultural capital that most popular forms of alterity relations – the fascination with new technology, the sacralization of certain brands, "escapist" forms of gaming, and so on – are met with suspicion (see also Danielsson, 2014; Bengtsson, 2015b).

This way of handling media things, as markers of taste and lifestyle, proves that we cannot understand the fluctuations of material indispensability and adaptation merely through the lens of techno-social dependence. People's ordinary need for, or disgust with, certain media, regardless of the type of phenomenological relation they may represent, point beyond the realm of technics. The need for properties is certainly not the same thing as the need for technics, and sometimes this leads to experiences of ambiguity among social people. This can be illustrated by means of another interview example:

> I've had an iPhone since last Christmas. I held out for a long time, I was wait-ing for my old Nokia to break but it never did. You just discover new uses for it everyday, I'd be lost without it. I don't have a great number of apps but Travel, Dictionary, Wordfeud, messages, email.
>
> *65-year-old woman from Stockholm*

This woman describes how she "held out" and wanted to use her old mobile phone as long as it was still working, expressing a distinct moral (anti-materialist) attitude towards the value of properties that was informed by cultural rather than economic capital. Eventually she got herself a smartphone as she felt the need to stay in touch with her son (who helped her to decide) and other relatives around the world via social media. As a further consequence, she has established a growing number of international relationships using her smartphone and now finds it difficult to do without it. In this case, therefore, the process of appropriation has been rather stretched out and was grounded in the value of particular hermeneutic relations rather than in the symbolic value (alterity relation) of the device itself.

A parallel example is the declining market for traditional newspapers. In Scandinavian countries the daily broadsheet has held an extremely strong posi-tion for many decades, especially due to subscriptions, and has been a more or less indispensable part of many people's everyday (morning) rituals. Competition from other media, including online distribution, is now threatening this position in both cultural and economic terms. Readers are more or less forced to appropriate new technologies in order to get access to their favourite news source. This signifies a general shift in the movement of mediatization in which one relation of indispen-sability replaces another. The shape of this new era of immediacy (Tomlinson, 2007) in which news is expected to be available "at one's fingertips" is illustrated by the interview extract that opened this chapter. Disconnection from the world of news becomes more or less unthinkable, as revealed in the respondent's experience of feeling "left behind" after less than a day offline. However, to certain segments of the media market such a shift means much more than just the adaptation to a new form of hermeneutic technics. It also means, potentially, the *loss of a signifying prop-erty*, namely the classified and classifying marker of printed newspapers, including their value as particular brands.

When analysing the significance of properties from a Bourdieusian perspective we are thus able to grasp in a deeper and contextualized sense the phenomenologi-cal complexity of technological relations, and thus the dynamics of mediatization. The fact that certain groups are willing to defend their printed newspaper, for example, shows that there are alterity relations, such as the sacralization of print media, at play besides the hermeneutic value of news reporting. This, in turn, can be taken as an illustration of the internal tensions of the mediatization meta-process – an expression of resistance to (trans)mediatization linked to the cultural desire to maintain clear boundaries in terms of time, space and social relations. From this point of view, the integrated and system-dependent nature of transmedia technologies constitutes a threat to individual autonomy and established criteria of cultural quality (such as "originality" and "objectivity").

The introduction of digital transmedia devices (smartphones, cameras, laptops, and so on) has tended to break down such modern categories. These devices have the potential to establish a diverse array of relations and can be dynamically adapted to serve different functional needs. At the same time, the interconnectivity and open-ended flow of digital data between different devices are turning the material spaces of everyday life into integrated media environments in which one particular function or relation might be established via various access points. As Madianou and Miller (2012) argue, the question of *which media* to use for fulfilling *which social need* is not related to functionality alone, but also, and increasingly, to moral and cultural pre-dispositions in combination with situational conditions. When processes of media appropriation become more open-ended, so does the value of media as properties. In a material environment where there are (hypothetically) no longer any record collections, newspapers or books to put on display, the cultural value, and thus the indispensability, of various devices will follow to a greater extent from *how* they are put to use; that is, *how* they are embedded in textural relations.

Media textures

Analysing textures does not mean that we turn to an entirely new dimension of media things. Rather, reaching the third level of analysis means that we need to look at media things in their dual capacity as technics *and* properties, as the means for building certain world relations as well as the means for cultural classification. Studying textures also means looking at the ways in which media things become indispensable not merely because of their functional and symbolic capacities, but also because of *what they feel like* when they enter into *patterns of amalgamation* through social practice. Texture thus brings together the key ideas of a cultural materialist framework, which as Wise (2012: 160) argues, "is more about resonance than representation, about forms and substances brought into relation". To some extent we have already touched upon these issues. In Ihde's work there are overlapping arguments in his discussion of the lifeworld as a "technologically textured ecosystem" (Ihde, 1990: 3), as well as in his discussion of background relations. In Bourdieu's (1997/2000) analysis of bodily knowledge we find corresponding observations as to how the positionings and relations of people and properties in social and physical space are both enacted by and inscribed in the body as a sort of ongoing material socialization and/or social materialization (see also Wiley et al., 2012).

More significantly, however, my understanding of texture builds on Lefebvre's (1974/1991) critical theory of the production of space. Here the concept of texture points to the "communicative fabric of space" (Jansson, 2006, 2007c), established through the meaningful repetition of spatial practices and ordering of communicative properties in space, and to the naturalized bodily and sensory experience, the "feel", of this fabric (see also Adams et al., 2001). Spatial practices are sometimes of a deliberately communicative nature, such as dinner conversations around the kitchen table or crowds of people gathering at the movie theatre in the evening. But they also include the infrastructure and everyday streams of activity that leave meaningful, communicative traces in social space: daily commuting patterns in the city, the spatial organization of our home environments, border arrangements at airports, and so on. All such arrangements are communicative.

Textures enable and give shape to certain types of communication in a given setting while excluding other types of communication (as well as groups of people). They thus support our sense of continuity and belonging (or "out-of-place-ness") both at the representational level and in an embodied sense as we learn how to move and act in various environments (Moores and Metykova, 2010; Moores, 2012). Accordingly, textures do not appear at random; they materialize through certain spatial and temporal regularities and rhythms:

> Paths are more important than the traffic they bear, because they are what endures in the form of the reticular patterns left by animals, both wild and domestic, and by people (in and around the houses of village or small town, as in the town's immediate environs). Always distinct and clearly indicated, such traces embody the "values" assigned to particular routes: danger, safety, waiting, promise. This graphic aspect, which was obviously not apparent to the original "actors" but which becomes quite clear with the aid of modern-day cartography, has more in common with a spider's web than with a drawing or plan. Could it be called a text, or a message? Possibly, but the analogy would serve no particularly useful purpose, and it would make more sense to speak of texture rather than of texts in this connection. […] Time and space are not separable within a texture so conceived: space implies time, and vice versa.
>
> *Lefebvre, 1974/1991: 118*

The last point helps us to further explicate the nature of indispensability. When theorizing how media things and associated media practices amalgamate with other spatial practices (Schulz, 2004) we may distinguish the *spatial/vertical* dimension from the *temporal/horizontal*.

Along the vertical dimension, amalgamation manifests as layerings or "thickenings" (cf. Hepp, 2009) of practices and artefacts at a particular place. This refers to the fact that social actors learn what to expect from certain places, and so shape places in terms of "what goes with what". In many institutional settings there are functional reasons for this type of amalgamation. In a train station, for example, travellers expect to find electronic information screens, timetables, clocks and surveillance cameras; these are some of the pre-conditions for the provision of efficient and safe transit systems. As illustrated above, even though the systemic imposition of abstract technologies, notably surveillance cameras, is not always socially embraced – depending on the type of setting and cultural context (see Jansson, 2012) – these systems are part of constructing the need for textural adaptation and routinization on behalf of social agents.

There are also spatial amalgamations based on cultural conventions and ritual practice. Many people today have the habit of reading the newspaper, checking out Facebook, or playing Wordfeud while waiting for or travelling by public transport. Certain media forms, translated into practice, thus have a stronger potential to amalgamate with certain spatial practices than others. Through repetition these amalgamations are turned into durable textures, implying that we "cannot have one thing without the other": "no running without my portable music", "no breakfast without my newspaper", and so on. The indispensability of a media device can here be traced to the fact that the overall feel of texture, the "comfort of things" that Miller (2008) speaks about, is ruptured, and associated practices even disabled, if the particular device is somehow missing or displaced. The indispensability of media becomes symbiotically linked to the normalization of social practices, thus reinforcing the overall mediatization of social space.

Along the temporal/horizontal dimension we find textural amalgamations grounded in routinized, or functionally interdependent, sequences of practice. We can express this type of temporal ordering in such statements as: "After doing this, I have to do that", or "Before doing that, I have to do this". Horizontal amalgamations thus create certain rhythms in everyday life and may take on different shapes in different cultural and historical settings. In an agrarian society, for example, we can envision the regular, mostly cyclic, sequences related to the cultivation of land and livestock. During industrialization, however, the integration of media technologies both took off and had an accelerating effect, which among other things demanded more abstract forms of time-keeping (Schivelbusch, 1987; Kern, 2003). Clocks and other time-keeping devices have had a pervasive effect on modern life, including in the private realm; even the very adjustment and maintenance of such technologies has been an amalgamated part of everyday textures (such as winding up the clock in the morning, or adjusting the alarm clock before going to sleep).

Perhaps even more prominently, however, the time-binding role of the media has been associated with broadcasting and the scheduling of radio and television programming. Examples range from the ritualized forms of Friday night gatherings in front of the television set to the more practical necessity of listening to weather forecasts before setting out on journeys in the mountains or on water. Foregoing such media practices, or the technologies that are indispensable for them, may evoke feelings of insecurity as well as emptiness.

Still, we must keep in mind Lefebvre's basic point that time and space are impossible to keep apart in actual processes of texturation. Horizontal amalgamation most often implies vertical amalgamation, and vice versa, since a particular (mediatized) practice will tend to occur at a certain time *and* place according to certain – functionally or culturally conditioned – logics. The textural inseparability of time and space testifies to the strength of certain amalgamations of media, the fact that particular technologies (often by way of the contents they carry) are felt to be indispensable at a certain *time and place*. Again, we can use an interview extract to illustrate the point:

Respondent: In the morning I usually read *Dagens Nyheter* or *Svenska Dagbladet* [Swedish newspapers], I switch, I usually go on the computer at work. I look at things that interest me, everything from booking training or recipes, mostly finding out things, looking for information. On the way home from work I usually read the evening *Aftonbladet* on my phone.

Interviewer: What about TV?

Respondent: Very little, just the few things I'm interested in. I can use SVT Play [Swedish Television's online media player] if there's something I want to watch. Also Metro on the metro.

45-year-old woman from Stockholm

This extract illustrates how mass media and transmedia are often intertwined in everyday textures. It is difficult to say whether the current shift from mass media textures to transmedia textures generates "stronger" or "weaker" amalgamations. The best that can be said is that they are *qualitatively* different, that indispensability evolves in partly new ways. This is reflected in the shift in people's expectations in terms of the *when and where* they expect certain media devices and information flows to be available and for what purposes. In the era of mass media, most access points were temporally and/or spatially fixed and pre-defined. Newspapers were categorized as "evening papers" or "morning papers" and distributed according to institutionalized transport routes either to specifically assigned media outlets or to the customer. For most of their history radio and television sets were, by and large, stationary technologies – everyday fixtures – whose contents were not easily trans-ferrable to other platforms (Moores, 1988; Adams, 1992). Audio- and video-cassette recorders, as well as portable devices, introduced some degree of flexibility (such as "time-shifting"), but it was not until the expansion of converging digital media discussed above that we could discern a major textural shift.

Above all, "transmediatization" means that media amalgamation with other practices becomes more open-ended and individualized. When media contents are expected to be available anytime and anywhere, and via different platforms, textures are no longer institutionally determined to the same extent. This does not mean that the material force of mediatization has weakened, however. As has been shown repeatedly throughout this chapter the versatility of transmedia devices enables them to be interwoven with everyday textures in increasingly complex ways. Sometimes this involves amalgamation with stable routines, such as regular Spotify listening in the car every morning. At other times, as Soukup argues in an ethnographic account of postmodern media culture, everyday life tends to be characterized by "fleeting moments without clear unity or sequence" marked by "the experience of being between screens and/or cultures rather than firmly entrenched in a single machine or cultural boundary" (2012: 234–35). What emerges is thus an increasingly ambiguous face of mediatization.

Conclusion: Three forms of media dependence

In this chapter I have further elaborated on the cultural materialist approach introduced in Chapter 2. The primary aim has been to introduce a conceptual framework for explicating how different technological, cultural and social forces interact in the shaping of mediatization, or, to put it differently, for distinguishing the principal (albeit interwoven) shapes in which media materialize as indispensable cultural forms. The framework includes a three-level approach based on the combination of Ihde's (1990) core notion of "I-technology-world" relations (*media technics*), Bourdieu's (1979/1984) sociological theories of socio-cultural legitimation and practical knowledge (*media properties*), and Lefebvre's (1974/1991) phenomenology of the materialization of everyday life *(media textures)*. As I have shown in historical and contemporary examples, media become normalized within ordinary life, taken up as part of practical consciousness, in the shape of technics, properties and textures.

We can speak therefore of the *triple articulation of media as cultural form*. While certain technological shifts, such as extended portability and simplified iconographic interfaces, may indeed contribute to the "disappearance" or naturalization of media within the lifeworld – and thus to the "lubrication of mediatization" – the full potential of such innovations of *technics* can only be realized so long as their affordances resonate with pre-established structures of practical knowledge and legitimation within concrete *settings of appropriation* and if the practical usage of new media devices creates strong *textural amalgamations* with various other social practices·in time and space.

The three-fold approach suggested here is thus instructive for identifying the internal contradictions and dialectical fluctuations of mediatization. When media get interwoven with everyday practices, enhanced opportunities for expressivity and connectivity are paired with media dependences, weak or strong, that emerge from overlapping orders of technology and recognition (see Figure 1.1). Even though media are always technics, properties and textural entities at the same time, it is possible to outline three ideal types of media dependence based on these articulations. The first type we may call *functional media dependence*, which is primarily tied to media technics. This type of dependence occurs when practical procedures are altered and made dependent on mediated forms of communication to the extent that a certain activity can no longer be carried out without the assistance of media. For example, the very constitution of many of today's ordinary spaces, such as modern supermarkets, rely on software-monitored processes. If the computerized infrastructure for making purchases in a supermarket were to crash, shopping would be impossible (since staff can no longer process goods manually) and the supermarket – ultimately defined as a code/space (see Kitchin and Dodge (2011) – would cease to be a supermarket.

Second, there is what we may call *transactional media dependence*, which is primarily related to media as properties. It refers to a set of conditions in which social actors comply with the rules and regulations set up by the media environment, including technological, institutional or cultural pressures, in order to achieve a certain form of gratification. This is typically the case when we speak of the mediatization of politics; that is, when politicians give up some of their autonomy in exchange for greater media exposure (e.g., Schulz, 2004). It is also a common feature of today's commercialized forms of surveillance, meaning that individuals sacrifice a certain amount of their privacy, giving away personal information, in order to take advantage of various online service benefits (e.g., Andrejevic, 2014).

More embedded forms of transactional dependence can be found in relation to communicational doxa and other areas of cultural normalization where social agents are (most often implicitly) forced to adapt their modes of communication to prescribed conventions of a certain social field or cultural setting in order to achieve recognition (see Chapter 2). It means that the appropriation of certain media is seen as mandatory within a given socio-cultural setting and that certain ways of using media become normalized as "standard procedure". We can think of the routinized ways in which many people use Facebook and other social media for sharing information with their peers, or the regular use of PowerPoint slides in work-related presentations. In most cases, none of these forms of communication is formally imposed or defined as mandatory in functional terms; rather, they can be seen as adaptations to social expectations and the successively evolving cultural order of things. The term "transaction" should not be taken in an instrumentalist sense, therefore; as Bourdieu (1979/1984) points out, cultural adaptations occur as the outcome of the "magical" fit between pre-dispositions, habitus and structurally prescribed modes of practice.

Finally, there is *ritual media dependence*, corresponding to media textures. This type of dependence is particularly difficult to delimit since it is based on the cultural and material power of shared routines and thus closely entangled with transactional dependences related to media properties. The key to understanding ritual dependence is to think about the power of habits, and what Bourdieu (1972/1977: 93) calls bodily *hexis*, referring to the continual inscriptions of culture into "durable manners" of doing things. As discussed above, the amalgamation of media within the regularized rhythms and spatial arrangements of everyday life, whether we speak of social television rituals or individual backstage media behaviours, imposes practical expectations on certain forms of activity at certain times and in certain places, which are stored in the body as a memory: "The principles em-bodied[sic] in this way are placed beyond the grasp of consciousness, and hence cannot be touched by voluntary, deliberate transformation, cannot even be made explicit" (ibid.: 94). Similarly, Lefebvre (1974/1991) underlines the power of habit and embodied practice in his writings on texture; it is precisely through such forms of patterned activity that textures are produced, maintained and ascribed meaning and power.

These three types of media dependence constitute a hierarchical order of indispensability, functional dependence being the most cohesive form.

As already mentioned, in real-life settings they are often interwoven. For example, ritual dependences often involve moments of transaction, and functional dependences are more often than not bound up with particular rituals and/or routines. The force of mediatization is at its strongest when dependences are normalized through the mutual reinforcement of media technics, properties and textures. This again underlines that the media alone do not create dependences. There is a plethora of social and cultural conditions that ultimately define what media should be used for and in what ways.

A further aim of this chapter has been to illuminate the fluid shape of mediatization and thus to position the book in relation to ongoing media change. After all, mediatization is a concept that, in spite of its meta-processual and relatively socially oriented character, aims to capture how changes in society are related to media developments. One cannot deny that the general *appearance* of mediatization is linked largely to the ways in which media work, to what types of communication existing technologies enable and what types of communicative needs they satisfy in certain contexts. Here I have tried to outline the implications of digital transmedia technologies and the emergence of transmedia textures in terms of a qualitative shift *within* the movement of mediatization. This does not mean that the fundamental meaning of mediatization, referring to its inherently dialectical nature, is in question. What changes with transmedia are *the ways* in which various conditions of media dependence unfold. Mediatization thus tends to *look* and *feel* considerably different in transmedia environments than in mass media environments, meaning for instance that new media amalgamate with pre-existing cultures and materialities in increasingly flexible and open-ended ways. In most social contexts, however, overlaps between these principal types of media environments can be seen as the rule, not the exception. We must therefore consider mediatization not just in relation to altered orders of technology, but also in relation to variations tied to social space and altered orders of social recognition. This is the focus of Chapter 4.

4
SOCIAL RECOGNITION AND STATUS IN A MEDIATIZED WORLD

In the previous chapter I discussed mediatization primarily in relation to changing orders of technology, trying to identify what characterizes mediatization in times of expanding transmedia environments. In this chapter I broaden the cultural materialist perspective and put the accent on cultural and social conditions rather than materiality. As already asserted, the dialectic of mediatization cannot be grasped unless we take into account broader (trans)formations of culture and society and see how such conditions correspond to the proliferation of certain technologies in everyday life; that is, how they develop into cultural forms. This was the basic point advanced by Raymond Williams (1974) and the reason why he compared the growth of television in the United States and the United Kingdom. In a similar manner I want to discuss the current face of mediatization, referring to its appearance in everyday life in relation to recent and ongoing changes within dominant orders of recognition. This means, broadly speaking, that I want to show how mediatization expands in tandem with the individualization meta-process.

Several researchers have identified the connections between mediatization and individualization (see, e.g., Hepp, 2013; Krotz, 2014; Christensen and Jansson, 2015). However, very few have explicitly linked mediatization to recognition theory. One exception is Stig Hjarvard (2013), who in his chapter on "the mediatization of habitus" sees the mediatization of modern society as a driving force behind the emergence of "soft individualism". Through mediatization, Hjarvard argues, "a paradoxical combination of individualism and sensibility toward the outside world has gained ground" (ibid.: 137). Hjarvard refers primarily to David Riesman's writings on the "other-directed" social character and consumerist lifestyles that emerged in the post-war United States, but also to Axel Honneth's theories of recognition and Pierre Bourdieu's concept of habitus. Similarly, Nick Couldry (2012: 171–73) discusses the relationship between media culture and recognition needs, pointing especially to the role of alternative and community media. Otherwise, in recognition theory there is a strong political and social philosophical bias (see Bankovsky and Le Goff, 2012) and it has gained relatively little attention within media and communication studies. In a recent volume entitled *Recognition Theory as Social Research* (O'Neill and Smith, 2012), in spite of the broad scope of the book none of the eleven chapters addresses the role of media in shaping contemporary orders of recognition. Probably the most extensive work to bring together questions of media and recognition so far is Luc Boltanski's (1996/1999) book on *Distant Suffering*. However, Boltanski's work deals chiefly with mediated spectatorship and is linked to questions of pity and self-justification in the context of mass-mediated humanitarian spectacles rather than to those questions of culture and everyday life that I want to explore in this book.

There are two inter-related reasons why I introduce recognition theory in the context of mediatization. First, as argued in Chapter 1, recognition holds a key position vis-à-vis the dialectic of mediatization. Social recognition is to be understood as a pre-condition for the development of an autonomous self and is thus a counterweight to media dependence. At the same time, as pointed out by Hjarvard (2013), individualized orders of recognition generate demands for increasingly reflexive lifestyles that rely on the mediated circulation of cultural values, norms and aesthetic styles. The social and existential ambiguities of mediatization, as seen for instance in the emergence of new structures of feeling, are thus intimately linked to the social struggle for recognition and to the coming of what Honneth (2004) calls "organized self-realization". Accordingly, and secondly, through recognition theory we can connect the cultural materialist perspective to an analysis of power relations in social space. The structural implications of "recognition work" are particularly salient in Bourdieu's (1979/1984) analyses of habitus, social status and modes of cultural distinction and legitimation. As McNay (2008: 11) argues, Bourdieu's idea of habitus "yields a notion of embodied subjectivity which is, in many respects, similar to the dialogical conception proposed by thinkers of recognition but is located more securely within a sociological account of power". Using Bourdieu's theorization we can thus explicate why certain groups in society, notably the mobile middle classes, are particularly inclined to become preoccupied with reflexive identity projects and thus also to emerge as key agents of mediatization.

The chapter begins with an introduction to Honneth's theory of recognition in general and his notion of organized self-realization in particular, and includes a discussion of how changing orders of recognition shape mediatization. I then offer a problematized view of Honneth's theory in which I discuss the role and challenges of social recognition in an increasingly mobile and mediatized society. I reflect specifically on the implications for hospitality, understood as the cosmopolitan ethos of recognition, in a world of blurred boundaries and intensified (trans) media flows. In the final section I connect recognition to questions of power by drawing on Bourdieu's cultural sociology. This means that I integrate recognition theory with the cultural materialist perspective, since recognition is always *also* a matter of classification involving (media) practices and (media) possessions as well as social agents. The discussion ends with an account of middle-class lifestyles, mobilities and trajectories. I argue, following the works of Riesman (1950/2001) and Bourdieu (1979/1984) as well as more recent studies, that recognition work finds its stronghold in the middle strata of social space, which is also where we find the epicentre of dominant culture. The middle classes normalize certain media (along with other goods and practices) as desirable properties and thus play a key role, hegemonic yet socially ambiguous, in mediatization processes.

Mediatization and organized self-realization

Human beings develop their identities through an ongoing interplay between social integration and separation. These dynamics mark social processes during the various stages of the life course in relation to changing social constellations. Recognition is a basic requirement for the individual to develop a sense of autonomous self; that is, to establish a sense of security in his or her own ability to think, reflect and act independently of other individuals. This sense of a secure "relation-to-self" (Honneth, 2012: 205) cannot emerge without positive attention from significant others, who contribute both to social integration and to the sense of self-worth on behalf of the individual. The individual's desire to belong to groups is thus not merely a reflection of integrative forces, but should be understood also as a quest for autonomy through recognition. One of the principles of Honneth's theory of recognition is that "groups should be understood, whatever their size or type, as a social mechanism that serves the interests or needs of the individual by helping him or her to achieve personal stability and growth" (ibid.: 203). However, membership of a group gives no guarantee of recognition in the true sense of the word, since groups may also involve repressive tendencies that lead rather to conformism and the dissolution of autonomy. Such problems regarding the integration of the "I" and the "we" can be seen as fundamental to social life from its very outset.

In Honneth's positive definition, recognition "should be understood as a genus comprising various forms of practical attitudes whose primary intention consists in a particular act of affirming another person or group" (ibid.: 80-81). The concept thus contains three basic premises: recognition should be (1) positively affirmative, (2) actualized through concrete action (rather than just symbolically), and (3) explicitly intended (rather than emerging as a social side effect or as the means for reaching other goals). It is also stated, as a fourth premise, that the basic attitude of recognition can take the form of various "sub-species", above all love, legal respect and esteem. Against such pure stances of recognition Honneth posits ideological forms that *exploit* the individual's psychosocial needs in order to instill attitudes that reproduce certain structures of domination. One example is the way in which societies of different epochs have assigned certain attributes to particular groups (defined in terms of, e.g., ethnicity or gender) as part of the structural reproduction of hegemonic orders for the assignation of work: "We could easily cite past examples that demonstrate just how often public displays of recognition merely serve to create and maintain an individual relation-to-self that is seamlessly integrated into a system based on the prevailing division of labour" (ibid.: 77). Such ideological forms of recognition are false, according to Honneth, because they fail to promote personal autonomy.

What makes Honneth's thinking around recognition especially interesting in the context of mediatization is that the concept lays the ground for a broader social criticism of modern society. Honneth's critique considers the ambiguous consequences of an extended individualization process, which in his view integrates forces, including the mass media, that under the auspices of supporting autonomy and recognition actually operate in the opposite direction. While the individualization process in its positive form promotes the growth of individual freedom and autonomy – setting individuals free from oppressive structures and normalizing the pluralization of choice – it has gradually (and especially since the last decades of the twentieth century) turned into another realm in which ideological forms of recognition prevail (Honneth, 2004, 2012: Ch. 9). There are today institutional expectations and ideological imperatives infusing a normalized view of "self-realization" as a required biographical goal for every individual. Media institutions as well as labour markets and a multitude of commercial actors encourage people to actively work on their "authentic self" and learn how to present their personality in ways that are as apposite as possible for reaching certain goals in society or in their careers. This organized form of self-realization implies that genuinely dialogical processes of recognition are undermined, replaced by standardized patterns of recognition and identity-seeking that merely serve the goal of legitimizing and further integrating individuals into the capitalist system. Authenticity and autonomy transmute into their opposites, simulation and conformism, and individuals may ultimately find their lives devoid of meaning.

What Honneth outlines is thus a dialectical transformation whereby the individual quest for recognition and autonomy leads instead in the direction of system legitimation and growing dependence (for similar diagnoses see also, e.g., Giddens, 1991; Boltanski and Chiapello, 1999/2007; Beck and Beck-Gernsheim, 2002). The role of the media is mentioned in only a few passages, such as this one: "Electronic media have certainly had a pioneering role in this process of redirection; their increased significance in everyday life now makes a much stronger contribution to sustaining the stylistic ideal of an original, creative life" (Honneth, 2012: 162). What Honneth seems to argue here is that "the media", taken as a compound institution, operates as a machine for normalizing desirable formats of self-realization, which in turn play the role of legitimizing certain ideological forms of recognition. This also means, if we assume that Honneth is right, that the transfiguration of individualization into organized self-realization is symbiotically interwoven with the dialectic of mediatization. The growing reliance on various processes of mediation for the gaining of recognition would imply that while "the media" contribute to the cultivation of false (or at least ambiguous) modes of recognition, such cultivation processes put into place an ideological structure that re-affirms the media as indispensable resources for human autonomy and growth, which would then, paradoxically, spur the process of further autonomy loss.

Beyond this important observation, however, Honneth's work does not offer any elaboration on the more precise status of "media", or on the ways in which they intervene in the dialectical expansion of organized self-realization. So how are we to understand the synergetic relationship between mediatization and organized self-realization and the dialectical relations that are produced? A key area to look into, I argue, concerns transformations of social space, the ways in which everyday lifeworlds and relationships between different social groups have become rearranged in modern society. The pluralization of both values and social milieus, or what Berger et al. (1973) call the "pluralization of lifeworlds", has led to a state of increased psychological vulnerability among individuals, which in turn can be seen as "*one*, if not *the*, central motive behind group formation today" (Honneth, 2012: 207). Furthermore, since modern society, as opposed to more traditional formations, does not provide one unified standard (such as religiously grounded ethics) in relation to which the individual may estimate the value of his or her achievements, it becomes increasingly important for each person to seek out the esteem of his or her peers. Here we can see how the prevalence of organized self-realization is rooted in broad transformations of social space in which place-bound communities (particularly those related to family and kinship) have loosened up and individuals have become increasingly mobile in their day-to-day activities as well as along their life trajectories.

This is undoubtedly a long-term transformation, which Honneth outlines partly via Simmel's (1900/1990) classic work on how urbanization and the monetarization of social relations affected individual autonomy and freedom in industrial society. A related diagnosis can be found in Riesman's (1950/2001) account of the modern shift from "inner-directedness" to "other-directedness" as the dominant mode of social conformity in post-war America. Riesman and his colleagues identified a growing social anxiety, especially among the urban, increasingly mobile middle classes, which they argued led to increasingly reflexive forms of identity and lifestyle management. The desire among peers to achieve mutual recognition was channelled through standardized consumption practices. Riesman paid considerable attention to the mass media, acknowledging their function as an agent for the circulation of standardized stylistic ensembles, a kind of omnipresent learning machine that could help individuals establish social bonds and channel their psychological needs for recognition in a spatially fragmented and volatile society. Yet another version of this perspective on modern mass media can be found in Lefebvre's critical analyses of popular culture and the "land of make-believe", examining for instance how advertising and magazines present consumer objects to audiences "with the codes that ritualize such 'messages' and make them available by *programming everyday life*" (Lefebvre, 1971/1984: 86, italics added).

Following these analyses, the mass media have operated, and still operate, simultaneously as a map and a guidebook of the social terrain, a representational system that establishes and negotiates the codes through which orders of recognition (and misrecognition) evolve. This means that the mass media not only mediate but also, and perhaps more essentially, *pre-mediate* the socio-spatial expectations and experiences of individual actors (Grusin, 2010; Jansson, 2013a), turning the process of (mass) mediation *as such* into a force of symbolic legitimation. As Couldry (2003a) suggests, the symbolic power of the media (taken in the broad, institutional sense) rests on a dominant mythology, or "programming" of everyday life (cf. Lefebvre above), that elevates the media to an institution that circulates symbolic material possessing exceptional social, cultural, economic or political significance. This mythology functions as a stabilizing factor in relation to the social uncertainties articulated through organized self-realization and other-directedness, and legitimizes the ritual dependence on the mass media as an order of "pre-mediated recognition" (leaving aside that these dependences take on contextually specific forms).

An important conclusion based on these observations is that mediatization is not a linear process, technologically or otherwise media-induced, but precisely the kind of meta-process that Krotz (2007, 2014) outlines. While changes in social space, involving altered orders of recognition, have been a force behind everyday media dependence throughout modernity, the capabilities of new media technics are contributing to the reshaping of these conditions. For instance, as discussed in Chapter 2, Williams (1974) introduced the term "mobile privatization" to show how ritualized uses of broadcasting technologies have underpinned the interplay between a home-centred suburban way of life and new forms of daily (auto)mobility patterns.

Today we must rethink these relationships. The widespread use of social media, mobile devices and numerous transmedia applications has expanded the social functions of mass media in recent years, both challenging and extending them. A growing share of media users, especially younger groups, orient their media habits towards interactive platforms, such as Facebook and YouTube, that circulate user-generated flows as well as content emanating from mass media industries. In Sweden, one of the leading countries in this development, more than 50 per cent of young Internet users (aged 12–18 years) access YouTube every day and more than 50 per cent of Internet users between 20 and 45 years of age watch YouTube every week (Findahl, 2014). As Gillespie (2010: 347) argues, these platforms have become the "curators of public discourse". They both enable and demand continuous monitoring and updating and thus feed off precisely those psychosocial needs and desires that characterize other-directed life environments while at the same time extending the pre-established mythology of institutionalized mediation as a marker of socio-cultural status. They normalize orders of recognition that differ in some respects from those implied by Honneth's discussion (which is of course partly due to the rapid development of new media), but it is obvious that the current stage of mediatization at a more fundamental level, and to a greater degree, feeds off precisely those psychosocial needs and desires that characterize the individualization process and thus contributes to the prolongation of organized self-realization. I will return to these questions later in the chapter. Before doing so, I wish to reflect some more on Honneth's theory of recognition, elaborating a view that responds to our increasingly mobile and complex society.

Recognition and hospitality in a mobile society

Honneth's theory of recognition exposes two problems related to his conception of identity. These problems stand out especially when his ideas are applied to conditions of growing cultural complexity. First, Honneth's model treats identity as a relatively homogenous entity and largely overlooks the often complex and internally contradictory relations-to-self that an individual may experience and express. His theoretical discussion ignores many of the ordinary problems that human beings have when they encounter one another in everyday life (or in institutional settings) and when ambiguities of identity (and thus the very object of recognition) are exposed and made the subject of interpretation. This is a problem that Honneth shares with, for example, Charles Taylor (1989), who in his theoretical programme on identity politics "of difference" tends to reproduce simplified group identities that do not match the multi-layeredness of real people's lives. The complexity and relative volatility of identity come to the fore particularly in relationships that develop between strangers and when individual subjects do not fit easily into obvious categories of, for example, ethnicity, gender or sexuality.

The second problem is more ethical in nature. Honneth advocates an *affirmative* model of recognition, which means that recognition should be based on the perception of qualities that other persons or groups already possess, rather than on the *attribution* of new (albeit positive) qualities. As he (2012: 81) points out, the problem of the attribution (or transformative) model is that it does not offer any "internal criteria for judging the correctness or appropriateness of such acts of ascription". Similar problems pertain to the affirmative model as well, but there, Honneth argues, the intersubjective potential of the lifeworld paves the way to reach valid forms of recognition. More qualified forms of value affirmation can be achieved through a progressive broadening of one's interpretative horizons:

> Without going into the details of such a process of progress, which I believe must be defined as a form of reflection on the knowledge that guides us in the lifeworld, the main idea behind it is that with the differentiation of evaluative qualities we observe and notice on the basis of our socialization, the normative level of our relations of recognition rises as well. With every value that we can affirm by an act of recognition, our opportunities for identifying with our abilities and attaining greater autonomy grow. This should suffice to justify the idea that our concept of recognition is anchored in a moderate form of value realism.
>
> *Honneth, 2012: 83*

What is agreeable in Honneth's approach is that he encourages a continuous learning process grounded in the ethical responsibility to actively pay attention to the qualities of others and thus also grow as an autonomous individual. Still, as Nancy Fraser has pointed out in several texts (see, e.g., Fraser, 1997: Ch. 1, 2000a, 2000b, 2001) there is a tendency in the affirmative approach to "encourage zero-sum thinking" instead of promoting synergy and transformation (2000a: 22). Affirmation cannot account for the complex dynamics of individual or group identities, but contributes rather to the reproduction of pre-established identity relations. The problem of affirmation then boils down to the basic theoretical flaw that was mentioned above; that is, the cultural reification of identity (Fraser, 2000b).

Any finite choice between affirmative and transformational models of recognition would be difficult to sustain in absolute terms; ultimately both models are based on theoretical suppositions and any attempts to fully validate them when confronted with real-life situations are bound to be problematic. Fraser's solution is to advance a model that moves beyond the ethical challenges of social interaction so as to sidestep the difficulties of the identity model of recognition altogether. In Fraser's (2001: 24) view, making recognition theory politically applicable would require that it emphasized the recognition not of group-specific identities but of "the status of group members as full partners in social interaction". This means that groups and individuals would be recognized as peers equal in status and in their ability to contribute to social life. Particular aspects of identity would then be protected from the risk of being misrecognized.

Fraser's perspective is indeed relevant at the political level; however, the obvious fact that she leaves ethical considerations aside (which also constitutes her main point) makes the status model non-applicable for comprehending and formulating a criticism of relationships of recognition. In day-to-day life the ethos of mutual learning that Honneth points to is not only an ethically desired stance, but is also grounded in those continuous cultural processes that make up the lifeworld.

My point is that the type of ethos that Honneth discusses does not have to be one-sidedly associated with the affirmative model of recognition, but may just as well have *mediating* between affirmative and transformative acts of recognition as its goal. In social interaction (between strangers or peers) this would signify an ethical stance in which openness to the possibility of mutual self-transformation is paired with a fundamental respect for what can be intersubjectively regarded as the other person's pre-established qualities. The ethos I am thinking of here can be seen as a certain form of *hospitality*, whose importance increases in a society of expanding mobilities and altered geo-social conditions. Under mobile conditions, when people more often encounter strangers as well as individuals or groups whose "qualities" (to speak with Honneth) are more or less unknown, *hospitality becomes the pre-condition for recognition*. I would even say that hospitality is the ethical *modus operandi* of recognition under mobile conditions.

The concept of hospitality, which has been widely debated in theories of cosmopolitanism, is just as complex as recognition, torn between the ethical extremes of affirmation and transformation. In Kant's (1795/1970) classic writings on hospitality as a universal law of the cosmopolitan world order (where all human beings have a similar right to the planet) he introduced a principle that granted the citizen who arrived in another state the right to visit; that is, to be openly welcomed as a guest, rather than met with hostility. In return, the host was granted the right of not being invaded or threatened in his or her territory. Kant's formulation (originally pertaining to the political level) is thus in keeping with an affirmative approach and pre-supposes that the guest is going back to his or her home country after the visit. The distinction between host and guest is clearly maintained. In a significant critique of Kant's perspective, Derrida (2000, 2001) has deconstructed the idea of hospitality and advocated a transformative approach that stresses the open-ended character of both host and guest. Hospitality as so conceived is a process of continuous negotiation in which the very act of welcoming becomes problematic because it pre-supposes somebody's right to claim a particular space as his or her own. Derrida thus suggests that the status of host and guest is in a state of constant flux, and that hospitality is always incipient, on the threshold (ultimately leading to the implosion of the concept itself).

In an attempt to reconcile these affirmative and transformative positions Mustafa Dikeç (2002) arrived at an understanding that converges with the intermediary, or mediating, conception that I advanced above:

> Thinking about hospitality, more importantly, is to think about *openings* and *recognition*. Although boundaries form an inherent part of the notion of hospitality, without which such a notion would perhaps be unnecessary, hospitality, I want to argue, is about opening, without abolishing, these boundaries, and *giving spaces* to the stranger where recognition on both sides would be possible. In this sense, it implies the *mutuality of recognition*.
>
> *Dikeç, 2002: 229, italics in original*

Dikeç's elaboration of hospitality is elegant in the way in which he, on the one hand, accounts for the dynamic nature of identity and the fact that encounters, especially between "strangers" (a term he discusses at length), necessarily involve some kind of change, at least in terms of those mutual learning processes that expand the individual's capacity for making future interpretations, and, on the other hand, takes seriously the deeper feelings of attachment and sense of home that individuals generally hold in relation to geo-social arrangement(s) of various kinds. As I read Dikeç's work, he represents a position that allows the cosmopolitan ethos to incorporate phenomenological understandings of place, home, identity and difference – such as those found in Tuan's (1977) and Seamon's (1979) work – without falling into spatial romanticism or cultural reification. The key is to envision hospitality as the *modus operandi* through which spaces are opened and kept open for processes of mutual recognition (and thus the growth of autonomy), which may then point to processes of self-transformation as well as cultural boundary negotiations and conflicts (which is not the same thing as misrecognition). Furthermore, hospitality takes time, we are reminded (Dikeç, 2009; Barnett, 2005). It is about *giving both time and space* to the stranger who suddenly stands on the threshold, rather than securing the border through which he or she is (re)produced as an Other (Dikeç, 2002: 244).

Where can we find such spaces of hospitality? How can they emerge and be kept open? And in what ways are they conditioned by our contemporary (trans) media environments? Much has been said and written about these issues, especially within the areas of urban studies and critical geography (see, e.g., Dikeç, 2009; Soja, 2010; Harvey, 2012), but also in research on media, migration and morality (see, e.g., Morley, 2000; Silverstone, 2007; Georgiou, 2013). The discussions have often been framed by the broader discourse on cosmopolitanism, cosmopolitanization and cosmopolitan culture (see, e.g., Beck, 2004/2006; Delanty, 2009; Papastergiadis, 2012). My aim here is neither to provide a comprehensive overview of potential answers to these questions, which would generate a wide range of political questions concerning social and spatial planning, nor to venture into the rather heated debates surrounding cosmopolitanism and its relevance as an ethical stance or epistemology (see Christensen and Jansson, 2015). Rather, I want to extract just one point from these discussions, which stands out as particularly important in relation to the proposed cultural materialist perspective of mediatization. This point concerns the key role of *spatial practice* and takes Dikeç's argument one step further while at the same time leading us back to Honneth and the dialectical view of mediatization.

The production of hospitable media spaces

Recognition, as Honneth suggests, cannot rely on merely symbolic activities, but needs to be anchored in concrete and consequential practices, which then also involves a certain moment of risk. The same thing applies to hospitality. To actively open up spaces to other people (literally or in cultural and emotional terms) is to initiate a process whose outcome is uncertain and in which each identity is potentially contested (Ivesen, 2006: 77; Silverstone, 2007). Derrida refers to this as the "double law" of hospitality: "to calculate the risks, yes, but without closing the door on the incalculable, that is, on the future and the foreigner" (Derrida, 2005: 6). The opening of spaces is thus to be regarded as a spatial practice, in Lefebvre's (1974/1991) sense of the term, that leads not only to the *sharing* of space in mutual recognition, but also to the very *production of space*. Hospitality invites people to work practically on the space: the guest who is invited into somebody's home does not leave that space totally unaffected, but contributes to its production, if not materially then in the sense that the home-place is (re)produced (at best) as a hospitable space to which those who live there can attribute a certain positive value.

Seen from the other side of the process, and as pointed out in the literature on urban justice and governance (e.g., Sandercock, 2000, 2003; Amin, 2002, 2012; Ivesen, 2006; Dikeç, 2009; Harvey, 2012), the concrete and collaborative production of space holds great potential to sustain mutual recognition among those involved. The city, understood as a pluralized space where "strangeness" is a condition potentially shared by everybody, provides enhanced opportunities for such interactions to occur (Ivesen, 2006). Still, it may require substantial political governance to actually bring about those micro-publics where strangers are forced to engage collaboratively in shared interests and "where dialogue and prosaic negotiations are compulsory" (Sandercock, 2003: 94). Using examples from various (trans)local projects of urban regeneration and development, Sandercock (2000, 2003) states that these processes become particularly valuable as part of the solution to xenophobia in that they recognize the need for "a language and a process of emotional involvement, of embodiment, of allowing the whole person to be present in negotiations and deliberations" (Sandercock, 2000: 26). In a related account, Amin (2002) describes "micro-publics of banal transgression" as spaces where people of different backgrounds come together, solve problems and create things. For example, he maintains that "Colleges of Further Education, usually located out of the residential areas which dominate the lives of the young people, are a *critical liminal or threshold space* between the habituation of home, school and neighbourhood on the one hand, and that of work, family, class and cultural group on the other hand" (ibid.: 14, italics added). Even the very *materialities* of space (locational properties, infrastructure, boundary arrangements, signage, etc.) work as unconscious mediators of cultural and moral value (Amin, 2012), thus making the cosmopolitan quest for open spaces and collaborative spatial practices an even more critical and forward-looking challenge. As I will show in Chapter 7, gentrification constitutes a type of spatial transition in which the tensions between transgressive, cosmopolitan ambitions and segregating forces are most clearly spelled out.

We can now discern how a spatialized understanding of hospitality establishes a link between recognition theory and cultural materialism, inviting us to think about media as tools for collaborative spatial production. Notions of digital media and online spaces as arenas for cultural experimentation and liminal identity exploration marked much of the prophetic discourses of the 1990s and are thus by no means new (see Mosco, 2004). Today's media research, however, involves a more historically grounded and increasingly significant turn towards materiality as a realm of cultural negotiation. While "media archaeology" is the most historically oriented articulation of this new materialism (see, e.g., Parikka, 2012), "collaborative media" is a branch that studies the expanded potential of digital media for realizing various forms of co-production and co-design of media contents, media spaces and *media as such* (Löwgren and Reimer, 2013). According to this orientation, collaborative media practices cover a broad spectrum, ranging from commercially interwoven processes channelled through social media platforms to more subversive and socially transformative projects associated with, for example, cultural governance, artistic interventions, or social movements and "tribes". The collaborative processes that Löwgren and Reimer identify based on a number of case studies converge in substantial ways with the vision of liminal, transgressive spaces outlined by urban theorists like Amin (2012) and Sandercock (2010). Collaborative media thus integrate a cosmopolitan potential, a potential to bring together relative strangers that has so far been exposed most prominently by alternative groups with identity, political or artistic agendas and interests in the nivellation of political and cultural hierarchies (see also Wilken, 2010; Christensen and Jansson, 2011; Papastergiadis, 2012).

The shift from mass media to collaborative transmedia thus denotes a promise of emancipation, recognition and autonomy through boundary-transcending spatial production in which the media are to be seen both as the tools for, and the very raw material of, such spatial production. However, from a critical sociological and human geographical standpoint the very terminology of "collaboration" stipulates a one-sided, and socially restricted, view of what current transformations of media environments actually do to social lifeworlds. The perspective of collaborative media puts the accent on just one, albeit important, side of the complex dialectic of mediatization. The other side, as I have already pointed out, is dependence. Beyond the kinds of transgressive cases that are discussed in certain literature on collaborative media there is ample evidence of profound social transformations in which the shift from mass media to transmedia (collaborative or not) assumes a logic of encapsulation (Jansson, 2007a, 2007b). This means, first of all, that media users are locked into infrastructural systems and hyper-surveilled spaces in order to connect and interact smoothly with one another, and, second, that the combined forces of organized self-realization and mediated social monitoring propel social life in the direction of further segregation rather than hospitality.

The first aspect, which we may call *techno-spatial encapsulation*, converges with the emergence of software-saturated spaces of code, or "code/space" (Kitchin and Dodge, 2011), that affect a number of everyday activities. While the implementation of digital technology infrastructure contributes to increasingly swift and frictionless mobility, which at the outset sustains the crossing of various administrative and cultural boundaries and thus the opening of spaces of encounter and learning, these expanded possibilities are premised on a basic trust in the justness and non-failure of these systems (cf. Giddens, 1991). As Amin and Thrift (2007) point out, spatial saturation with technology infrastructure is not a new phenomenon, but rather constitutes a historical development in which new layers of a taken-for-granted "machinic order" are continuously added to older ones (see also Graham and Marvin, 2001; Jansson, 2010b). In urban areas especially, and in relation to transport systems, such technological orders, which include objects as diverse as road signals, postcodes, pipes and overhead cables, satellites, office design and furniture, clocks, commuting patterns, computers and telephones, become pervasive:

> In the city, these objects are aligned and made to count through all manner of intermediaries such as rhythms of delivery or commuting, traffic-flow systems, integrated transport and logistics systems, internet protocols, rituals of civic and public conduct, family routines, and cultures of workplace or neighbourhood.
>
> *Amin and Thrift, 2007: 153–4*

The normalization of code/space, which is aligned with the development of trans-media systems, can thus be seen as the latest stage in a much longer sequence of techno-spatial encapsulation in which the growing ease of mobility and interaction operates in tandem with growing functional and transactional dependence. Today, the accelerating development of geographical information systems (GIS) and locative media, paired with mobile media devices for personal use, seems to propel this encapsulation to yet another level, anticipating the coming of an "Internet of things" in which geo-tagged objects in the environment become carriers of information (actually stored somewhere else) that ordinary people can also play a part in producing and circulating (e.g., Lapenta, 2011; Felgenhauer and Quade, 2012; Wilken and Goggin, 2015; McQuire, 2016). While these new affordances illustrate emerging possibilities for alternative and/or collaborative re-codings and re-orderings of space in a concrete way, quite literally abolishing the distinctions between mediation and spatial production, they also mean that ever greater shares of people's spatial and communicative practices are monitored (e.g., Andrejevic, 2007). By extension, as Crampton (1995) had already indicated in the early 1990s, these contradictory conditions raise a broadening set of ethical issues that are often difficult to resolve.

The second aspect points to the socially segregating logic according to which these developments tend to unfold; that is, *geo-social encapsulation*. This logic has a more direct impact on the production of open-ended spaces of hospitality and recognition. In simplistic terms, it refers to the fact that transmedia spaces tend to sustain people's endeavours to nurture pre-existing social networks and communities, moving about without risk rather than engaging in foreign cultural spaces, ideas and subjects. There are both institutional and social drivers behind this. On the one hand, there are commercial and political–administrative interests in governing consumers and citizens in their spatial practice. This is seen in diverse areas, ranging from interactively and algorithmically generated online advertising that encourages consumers to stay within their reproductive enclaves of preference, or what Pariser (2011) calls the "filter bubble" (see also Morris, 2015; Striphas, 2015), to the social sorting of citizens through automated forms of boundary maintenance and profiling that make mobility and access to certain spaces more complicated for certain groups than for others (Graham, 2004, 2005; Parks, 2007; Lyon, 2007).

On the other hand, the prevalence of other-directedness, the desire to be symbolically affirmed by one's peers and the ideological imperative of organized self-realization give shape to more or less insular flows of communication and bounded solidarities (Ling, 2008). As Turkle (2011) argues in a critical account of our increasingly connected lives, even in public spaces (of transit, consumption, leisure, and so forth) individuals are absorbed into mediated worlds and attend to the goings-on of geographically distant peers. As these forms of "connected presence" (Licoppe, 2004), or what alternatively could be called "mediated absence", grow stronger it becomes less likely that individuals will run into problematic encounters with others or have to deal with complex issues of cultural and/or emotional negotiation in face-to-face situations.

There is a danger that the mutual reinforcement of new orders of technology and new orders of recognition are generating encapsulated life-conditions, contradicting the basic principles of hospitality while at the same time making media connectivity indispensable to social life (see also Abe, 2009; Molz, 2007, 2012, 2014; Striphas, 2015). This can be seen above all in relation to dominant social, or connective, media – by which I mean social networking sites (e.g., Facebook, LinkedIn), video sharing sites (e.g., YouTube), blogs and micro-blogs (e.g., Twitter, Weibo), as well as social media extensions of various lifestyle applications (e.g., RunKeeper, Nike+) that turn "platformed sociality" into economic value through the implementation of industrial logics (Van Dijck, 2013: 4). These industries build their success largely on the promise of providing solutions to recognition deficits, but contribute at the same time to the reinforcement of a culture of connectivity through the circulation of simulated forms of recognition. Even though different platforms encourage different forms of interaction and exposure, reproducing certain tastes and orders of recognition (see Papacharissi, 2009), the architecture of these media and the interfaces through which communication unfolds sustain open-ended processes of simulation in which the distinction between connectivity and connectedness is collapsed. For instance, while algorithmic systems may keep track of how many connections (friends, followers, etc.) different users have and how many confirmative acts certain posts generate, these functionalities contradict the dialogical elements that mark pure forms of recognition and make it possible for each communicator to hermeneutically assess and build trust in the intentionality and practical relevance of the symbolic acts of others (e.g., Van Dijck, 2013; Striphas, 2015). On the contrary, social media relations are typically marked by *uncertainty* as to what intentions and what level of involvement may hide behind the digital interface. This mediated social uncertainty can be identified in areas as diverse as political action (e.g., related to micro-blogging) and intimate relations (e.g., dating sites).

This is not to say that all forms of interaction that occur via connective media resonate with the industrially invoked logics of popularity scores and simulations of connectedness or that all forms of recognition in these platforms are of an ideological nature. Nor is it to say that connective practices, such as liking, commenting and (geo-)tagging, cannot be part of deeper relations of recognition, such as love, friendship or identity politics, or make up community oriented flows of phatic communication (see, e.g., Miller, 2008; Ling, 2008). What I do suggest is that we are currently witnessing a qualitative shift pertaining to what mediatization *looks like* and *feels like*. In twenty years we have moved from relatively identifiable "spaces of media dependence" to increasingly pervasive "media spaces of dependence". As Couldry and McCarthy (2004) define the term, media space points to various forms of entanglements, on different scales, between spatial production and processes of mediation. For much of the twentieth century, and still today, modern life was spatially and temporally ordered in relation to the material and cultural properties of mass media (see, e.g., Spigel, 1992; Scannell, 1996). While broadcasting, taken as one institution, enabled new forms of social extension and functioned as a (pre-)mediator of recognition, as discussed above, it also established (more or less context-specific) dependences vis-à-vis certain flows of information and certain technologies. The coming of transmedia means that media users today are generative agents embedded within the flows they themselves consume, while also being dragged into other processes of mediation through their everyday practices of, for example, consumption and mobility. Accordingly, there is no longer any easy way of "opting out" of media space. The boundaries between offline and online, between the outside and the inside of mediation, are dissolving. This shift can only be understood if we account for how changes of media, such as the new industrial logics of connective media, resonate with social forces already at play in individualized societies, above all the increasingly dominant order of organized self-realization.

To sum up, the counter-argument, or sociological corrective to collaborative media is that the new affordances of media endorse developments that disintegrate society and make people increasingly absorbed in their own socially and culturally homogenous capsules, whether locally anchored or dispersed in space. In these new media spaces even expressions of hospitality run the risk of obscuring or standing in the way of practical acts of recognition, getting mixed up with calculated acts of self-branding and moral legitimation. In a future "Internet of things" where non-human "infomediaries" will ensure that we no longer have to reflect on what we like or where we should go to experience the things we desire (Morris, 2015), we may hardly have to encounter anything that forces us to reconsider our own orientations. Set against such an anti-cosmopolitan scenario, critical mediatization research has an important role in sorting out how the dialectical relations between autonomy/recognition and dependence/encapsulation are played out in different cultural contexts and in different parts of social space.

The cultural ambiguities of mobile middle-class identities

So far in this chapter I have discussed how altered media dependences both respond to and legitimize the ideological imperatives of organized self-realization. This expanding order of social recognition can be seen as an expression of dominant culture, the dominant direction that the individualization meta-process takes in contemporary capitalist society. Transmedia in general, and connective media in particular, reproduce this force at the most mundane level of ordinary culture, influencing the recognition work of people at basically all stages of life and in most social contexts. People can seek out mediated affirmations of their identity projects and make comparisons with the lives of others (mainly people within their peer groups) more or less regardless of time and place, practices that feed into the industrial aggregation of valuable information and algorithmically governed production of lifestyle simulations. In extreme cases, and as the ultimate expression of organized self-realization, individuals start conceiving of themselves as "brands" and use social media as a fluid and public stage for self-promotion (see Marwick, 2013).

In relation to such developments, and even though I have identified organized self-realization as an overarching hegemonic force that moulds contemporary mediatization in a certain way, we need to assess what the actual conditions for mutual recognition look like in various realms of society and in relation to different media constellations. We are dealing with a highly complex set of transformations that do not look the same everywhere in the world or in all parts of society. In this book I concentrate specifically on *mobile lives*; that is, *the structure of feeling associated with mobile middle-class lifestyles and life trajectories*. As explained in Chapter 1, this is not a random choice or a choice of convenience. The underlying rationale is, firstly, that mobile lives constitute a site where the dialectic of mediatization, the interplay between emancipation and dependence, are most clearly spelled out. Second, the middle classes constitute the epicentres of dominant culture, which means that they have a formative function in the negotiation and normalization of media in society. Through analyses of the mobile middle classes, I argue, we can learn something about the general directions that mediatization is currently taking. But how can we substantiate these vantage points? And what are the main traits of mobile lives as a structure of feeling?

To begin with, we should recall the general connection between the middle classes and reflexive identity work, or *recognition work*. As I discussed above, there are parallels between Honneth's (2004) view of organized self-realization and Riesman's (1950/2001) other-directedness, a mode of social conformity identified chiefly among urban middle-class groups. According to Riesman, the other-directed personality, whose sense of individual continuity and meaning is derived from affirmative recognition within the peer group, is an outcome of general transformations of modern society, especially the weakening role of traditional structures and increasingly pluralized life-conditions (see also Berger et al., 1973). The middle classes are thus affected in a double sense, and are also *produced*, by the social transformations of modern society. They can be found in urban/metropolitan districts and within professional sectors that demand reflexivity due to high degrees of differentiation and change, *and* they tend to be socially mobile, driven by what Luckmann and Berger (1964) term a *mobility ethos*. This has also been shown in subsequent analyses of new middle-class groupings, such as the "new cultural intermediaries" (Featherstone, 1991; see also Bourdieu, 1979/1984) and the "creative class" (Florida, 2002), where the maintenance of weak ties and open-ended networks gains prominence over more durable communities (Wittel, 2001). Under such life-conditions continuous social adaptation and identity management become pre-requisites not just for social recognition in general but for the appropriation of status positions in particular, which means, by extension, that there are great risks of anxiety, dissonance and disappointment.

There are numerous studies testifying to this general trait of the middle classes, and indeed there are several facets to it. The most comprehensive view by far is provided by Bourdieu (1979/1984), who in *Distinction* demonstrates how lifestyles and cultural tastes (re)produce social status positions by means of mutual processes of classification. Aspiring to a certain position implies that there is a willingness and ability on behalf of the social agent to appropriate broadly recognized forms of legitimate culture (including objects as well as practices). While those who belong to more privileged social classes and have inherited large amounts of cultural capital tend to develop "good taste" in a seemingly natural way, it is more complicated for those who are socially mobile and forced to rework their habitus through continuous cultural accommodation. Bourdieu stresses that middle-class subjects are inclined to recognize legitimate forms of culture as desirable – referring to art, cultural products and lifestyle practices associated with the dominant classes – thus testifying to their *aspirations* for upward social mobility. At the same time they are less capable of actually *knowing* and *appreciating* these markers of distinction in a relaxed and more elaborated way.

> Uncertain of their classifications, divided between the tastes they incline to and the tastes they aspire to, the petit bourgeois are condemned to disparate choices (which the new petit bourgeoisie, with its concern to rehabilitate folklore and exotic music, actively pursues); and this is seen as much in their preferences in music or painting as in their everyday choices. […] What makes middle-brow culture is the middle-class relation to culture – mistaken identity, misplaced belief, allodoxia.
>
> *Bourdieu, 1979/1984: 326–7*

The underlying pre-requisite of this statement is that we should think of the middle classes, and classes in general, as a *relational construct* rather than as a fixed set of "objective" positions. Classes are continuously (re)produced and negotiated through cultural and political struggle. *Allodoxia* is an important term for understanding these processes, invented by Bourdieu as way of conceptualizing the mismatch between social habitus and dominant structures of classification, articulated as cultural mis-identifications and mis-interpretations among socially mobile, or displaced, subjects. It means that these subjects often fail to feel at ease in the positions they may have formally appropriated with their educational and professional achievements. It also means that they must distance themselves to some extent from the forms of culture they are most attuned to through habitus in order to avoid experiences of de-classification.

The idea of allodoxia is thus closely related to the above discussion about affirmative versus transformative modes of recognition and shows how mobile middle-class identities (more than others) are marked by ongoing conflicts between these forms. Mobile subjects are torn between a search for transformative recognition pointing in the direction of their social ambitions and a desire to have their existing habitus affirmed rather than de-classified. This general *cross-pressure* (Blau, 1994: 41) translates into a variety of cultural contradictions pertaining to middle-class tastes.

Upwardly mobile subjects are often *cultural omnivores*, which means that they are more likely to combine cultural elements from both popular culture and high-brow culture within their lifestyles. In the 1990s many studies pointed to the rise of the "new middle class" (Savage et al., 1992), or class fractions socialized within a "self-realizing milieu" (Schulze, 1995), whose lifestyles were characterized by omnivorous taste patterns (Peterson, 1992). Demographically, these "postmodern" groups were relatively young, well educated and upwardly mobile (see also Featherstone, 1991; Wynne and O'Connor, 1998; van Eijck, 1999). Even though some of the conclusions regarding these new class fractions have been further elaborated in later research, the relationship between upwardly social mobility and cultural omnivorousness, or dissonance, has been confirmed repeatedly in studies from different contexts (e.g., Emmison, 2003; Lahire, 2008; Savage and Gayo-Cal, 2011; Daenekindt and Roose, 2014). While the middle classes gradually adapt to new social milieus, searching for recognition from their new peers, they also develop strategies for legitimizing parts of their own cultural background, notably popular culture, as a means of cultural affirmation. This is most clearly expressed among the "new cultural intermediaries" and symbolic experts employed within the media and culture industries (see, e.g., Featherstone, 1991; Lash and Urry, 1994), where the divisions between "high" and "low", and between creative and traditional realms of culture (see Williams, 1961/1965), are actively contested.

A related expression of cross-pressure is the tension between *cultural openness and disgust*. The omnivore thesis is often associated with phenomena like cultural flexibility and cultural literacy (e.g., Lizardo, 2006; Johnson, 2014), even cosmopolitan openness and sociality (e.g., Salazar, 2010; Meuleman and Savage, 2013). Mobile individuals located in pluralistic lifeworlds are said to be particularly adroit at taking the other's perspective and adapting to new cultural environments. Such cosmopolitan skills are grounded in the mobility ethos and reinforced through middle-class, social and geographical, trajectories (see Christensen and Jansson, 2015). At the same time middle-class identities are constructed in relation to *what they are not*, notably vis-à-vis the tastes and lifestyles of the lower classes. As Lawler (2005) argues, the middle classes tend to think of themselves as the bearers of progressive, cosmopolitan attitudes, which means that there must be *others* who do not embrace such values. These others, notably the working classes, represent bad taste and the type of *commonness* that middle-class subjects need to distance themselves from – even though working-class culture may be part of their own backgrounds. Cultural openness is thus rarely the embodied ethos that Dikeç (2002) calls for, but rather a conditional attitude that works in directions that are beneficial to social mobility; that is, as a means for accumulating symbolic capital (Bourdieu, 1979/1984).

Lawler (2005) also points to the hegemonic reproduction of middle-class identities via the media. According to her analyses of British popular media there is a lack of reflexivity regarding the perspectives from which the social world is depicted. Whereas middle-class points of view are normalized – partly because media people are also middle-class people, partly because the middle classes constitute desirable consumer segments – the lives of working-class subjects are obliterated or treated with disgust. However, these acts of discrimination do not address the "objective" markers of working-class positions; rather, they are based on aesthetic judgements that, often implicitly, associate certain tastes with weak character and lack of originality. Accordingly, as Bourdieu (1979/1984: 467-70) also points out, middle-class identities are shaped in relation to an undifferentiated mass, whose "vulgar tastes" are sometimes met with fear, symbolizing the broader decline of society, while at the same time providing the necessary counterpart to self-legitimation. In this way, middle-class identities come to occupy an ambivalent and *irresolvable* in-between status in which the hegemonic character of middle-class lifestyles, as seen at the collective and representational level, is set against aspirational individual trajectories and life experiences marked by uncertainty, anxiety and incompleteness. The middle classes are thus continually involved in the mediation, through recognition and misrecognition, of all those antagonistic value judgements (high and low, spiritual and material, fine and coarse, light and heavy, free and forced, broad and narrow, unique and common) through which commonplace understandings of social structure take shape.

The overarching cross-pressures introduced here, and the different social and existential dilemmas that may follow, can be identified and analysed in multiple contexts of social life. As shown by the dialectical line of argument pursued in this book, there are particularly important questions attached to the fact that exactly those media affordances that empower individuals to become more mobile and pursue their self-realizing, potentially cosmopolitan, life trajectories are also those that pull people closer together within enclaves of the like-minded (Jansson, 2011, 2014; Polson, 2016). Molz (2007, 2014) gives an interesting example in her study of how the reputation systems of mobile online platforms such as Couch Surfing shape the practices of global travellers. These services enable travellers to get in touch with people around the world and find affordable accommodation in order to explore cultural difference in a safe and secure way. At the same time, the business logic of these services is based on the expectation of exchange, which contradicts the idea of hospitality and tends to promote socially homogenous spaces where only certain types of difference are encouraged. Similarly, Andersson (2013) shows in a study of voluntary and non-voluntary migrants (Swedes in the Netherlands vs. Bosnians in Sweden) that social networking sites like Facebook are regularly understood as mandatory parts of mobile life trajectories. These media enhance the spatial freedom of privileged groups who can keep track of friends and acquaintances around the world. They also assist most migrants to maintain a sense of their roots and stay in touch with people in their home countries. At the same time there is often a need to manage social media (i.e., through profile maintenance) in order to filter out contents and persons deemed undesirable because they may evoke feelings of biographical disruption. Such acts may exclude people and events from the past (in Andersson's study, related to memories from the Balkan wars) as well as certain contemporary phenomena that do not (or no longer) "fit" one's identity.

I will not provide any further examples at this point, but will return to this discussion in Chapters 5–7 where three distinctive, and inter-related, expressions of mobile lives are analysed: *elite cosmopolitanism* (expatriate lifestyles), *post-tourism* (urban exploration) and *gentrification*. What these forms of privileged geo-social mobility have in common is that they can be defined at the intersection of individualistic ambitions of self-realization and everyday struggles to belong and receive recognition. They are thus taken as empirical sites for grasping the deeper experiential layers of a broader structure of feeling, *mobile lives*. They also represent highly mediatized social phenomena whose broadly recognized value reproduces the hegemonic status of reflexive middle-class lifestyles in a mobile society. Longer stays abroad are often seen as mandatory parts of upwardly mobile trajectories, a kind of forced freedom, corresponding to popular images of cosmopolitan lifestyles (e.g., Salazar, 2011; Glick-Schiller and Salazar, 2013). Post-touristic identities represent the middle-class ambition to escape what touristic mobilities are generally associated with: the mass movement of people following standardized taste patterns (Urry, 1988; Munt, 1994). Gentrification processes are imbued with popular mythologies of the "authentic self" achieved through creative place-making practices and the exploration of difference (e.g., Savage, 2010). As my analyses will show, within all three realms the media constitute valuable, even indispensable, tools for self-realization – in their capacity both to shape hegemonic imagery of social and geographical mobility and to support the appropriation of new spaces, places and practices – while at the same time contributing to ambiguous experiences of dependence, restraint and social enclosure.

Conclusion: Bridging recognition theory and cultural materialism

In this chapter I have elaborated four major points related to how mediatization affects identity development in modern society. First, I argued that the dialectic of mediatization operate in tandem with broader societal transformations, notably individualization. Building on Honneth's theory of recognition I pointed to the growing significance of media in a society in which organized self-realization reigns and where social recognition, and thus individual autonomy, demand reflexive work. Second, I argued that recognition in mobile and culturally complex societies necessitates hospitality as its *modus operandi*. Critical analyses of mediatization should therefore assess whether and how our contemporary media environments sustain such processes, and under what conditions. Third, I argued that the rise of transmedia technologies involves a dual movement: while the collaborative production of transgressive media spaces may sustain mutual relations of recognition, dominant social media also invoke flawed processes of recognition and extended media dependence. Based on these arguments, I posited that mobile middle-class lifestyles constitute a particularly relevant research context for gaining deeper knowledge about the dialectic of mediatization.

On a theoretical level this also means that I have attempted to build a bridge between recognition theory and the cultural materialist perspective of mediatization. Through recognition theory we can develop a greater sensitivity to the social and psychological drivers behind the everyday appropriations of media – the need to feel part of a meaningful social environment. At the same time, the cultural materialist perspective, especially Bourdieu's (1979/1984) work on taste and symbolic struggle, helps us see how processes of recognition are grounded in the everyday world of classified objects and practices. When media are established as cultural forms, as discussed in Chapters 2 and 3, they also become increasingly important to people's sense of belonging and thus constitute a cultural-material foundation for recognition work. This concerns *what* media people use as well as *how* they use them and in what ways they are incorporated as properties within broader ensembles of classified cultural objects and practices. As Bourdieu (ibid.) shows, these continuous processes of (re-)classification build up a space of lifestyles where some properties are recognized as more valuable than others, which in turn leads to the stratification and ordering of social agents and their trajectories according to status.

This brings us back to the question of cultural power and the hegemonic structures of opportunity and restraint. Recognition, as we have seen, is more than just a matter of achieving a sense of social autonomy and continuous self-identity. It is also a mechanism that (re)produces power relations in society and everyday life and manages the dynamics of inclusion and exclusion. While certain groups have to struggle, *as groups*, in order to achieve a sense of dignity and respect vis-à-vis the rest of society, others are more privileged from the start. As I have argued in this chapter, the mobile middle classes occupy a rather ambiguous position in this regard. On the one hand, they possess the resources to be continually engaged in their own self-realization, understood as a matter of building self-esteem through various achievements (as part of their lifestyles) rather than gaining basic confidence or legal respect. The media are today indispensable ingredients for such projects of the Self. On the other hand, to the extent that these groups are mobile (socially, geographically and culturally) they are also vulnerable to being symbolically rejected as they move from one life context, or one state of being, to another. In order to be mobile they have to re-negotiate or break with the cultural restraints – classificatory structures, everyday textures, and so forth – that hold them in place and grant them affirmation. In short, mobile middle-class identities are stuck in the Hegelian Master/Slave dialectic and thus constitute a valid site for researching the ambiguous role of media in shaping ordinary culture in general and what I have called hospitable spaces in particular. These are the deeper questions of this book, to which I now turn.

PART II

Inside Mobile Lives

5
MEDIATIZATION AND ELITE COSMOPOLITANISM

> Working is like my interest. And now when we are working across several time zones I might know for instance that there is something happening in Latin America that I want to keep track of and then I can check it out online in the evening and see if there is something I need to tweet or just keep track of when I get back to work the next morning. Or if somebody is giving a talk at a conference over the weekend I can keep track of it during the weekend. This is something that nobody demands from me, but it's just that I think it's part of my job. Perhaps it's a "good-girl-thing", but I want to know what's going on.
>
> *Matilda – information officer in her forties*

Matilda works in Geneva at the headquarters of an international organization concerned with human rights. She comes from Sweden and worked in several countries before settling in the Swiss city with her husband and children. She holds an advanced-level university degree and has invested much of her identity in both her professional career and her transnational family. In recent years, especially before arriving in Geneva, she has put great effort into coordinating mobile work-conditions and an internationally distributed family life, including both longer and shorter periods of geographical separation. An underlying driving force that explains why Matilda's work is also her interest is her deep engagement with international development and global justice. New media technologies, for their part, have given her important tools for managing her complex life and further contributed to the erosion of boundaries between professional and private time-spaces. The combination of mobile life-conditions and an emotional dedication to both work and family can thus be seen as a key driver of mediatization.

In this chapter, following Bude and Dürrschmidt's (2010) call for an existential turn in globalization theory, I want to recognize not just the emancipatory potential but also the social and existential costs that media dependence actualizes among privileged and geographically mobile groups (see also Jansson, 2016). In line with the ideal of immanent critique, my aim is to capture the experiential ambiguities and ongoing negotiations that articulate mediatization at the level of everyday life. In doing so I pay particular attention to the expanding orders of transmedia technology (Chapter 3). As shown in previous research, the ubiquitous nature of contemporary media contributes to the normalization of "flexible work", enabling white-collar professionals to invest more of their time and energy in their careers while at the same time being able to entertain family life and close relationships. However, this new flexibility – which has often been celebrated as emancipatory most of all for women – also incorporates considerable elements of stress and anxiety due to the fact that personal relations have to be either managed at a distance or placed in the background of professional tasks (Gregg, 2008, 2011). Such everyday sacrifices stand in sharp contrast to the mythologized and discursively promoted view of the networked "frequent flyer" or "business class traveller" as the archetype of global status and prestige (see. e.g., Calhoun, 2002; Thurlow and Jaworski, 2006). I therefore suggest that expatriate middle-class lifestyles provide a particularly good entry-point for studying both the cultural cross-pressures of mobile lives and the dialectic of mediatization.

Against this background, this chapter offers an analysis and critical discussion of the role of media within the expatriate lifestyles of *elite cosmopolitans*. The label "elite cosmopolitans" refers to those super-mobile subjects who do not relocate to a foreign country for just one or two periods of their careers – which in many sectors of society has become an increasingly mandatory thing to do (see Polson, 2016) – but whose careers are inherently dependent on continuous geographical mobility between different postings and whose multi- and/or cross-cultural experiences and competences constitute a form of capital (see Igarashi and Saito, 2014). As I discuss below, this *does not mean* that the group constitutes an elite in socioeconomic terms. However, in a world of involuntary mass movements, on the one hand, and banal, mediated cosmopolitan experiences on the other, they occupy an extremely privileged position, relatively speaking (Szerszynski and Urry, 2002). And seen in relation to organized self-realization as an overarching order of recognition (see Chapter 4) their life trajectories symbolize a rather exceptional level of achievement.

Further, the term "elite cosmopolitan" *does not mean* that we study life trajectories that are cosmopolitan in the stricter, ethical sense of the term (see Delanty, 2009). Rather, it means that the fields in which these individuals are employed maintain and legitimize basic cosmopolitan values (such as global justice, cultural openness and hospitality) as key elements of doxa. It also means that these mobile subjects *see themselves* as cosmopolitans; that is, they actively maintain a cosmopolitan self-identity that is distinctly different from local, sedentary ways of life. Accordingly, as Thurlow and Jaworski (2006: 103) suggest, I approach "elite cosmopolitanism" as a structure of feeling that comes about through the enactment of a particular kind of socio-culturally and discursively prescribed mobile identity.

Empirically, the analysis revolves around the mobile lives of Scandinavian expatriates ("expats") residing in Geneva and working for international organizations that are part of or associated with the United Nations (UN) (see Chapter 1 for a presentation of the material). The rationale for looking more deeply into a confined socio-spatial setting is that it gives us the opportunity to work more systematically with Bourdieusian field theory, notably the concept of communicational doxa (see Chapter 2), and to come to terms with how the logics of the field are played out along with other cultural and material forces, such as place-specific infrastructure, habitus and gender structures, within the continuous (re)shaping of mediatization. It should be stressed that the main points are also applicable to the broader interest in mediatized and increasingly mobile professional careers in which the appropriation of new places is not primarily a matter of settlement and elective belonging but the construction of distinctive geo-social trajectories. My arguments align closely, for example, with Polson's (2016) insights regarding the expansion of privileged, borderless identities – ultimately a new global, mobile and connected middle class.

The chapter begins with a general overview of how cosmopolitan self-identities are related to, and ultimately depend on, various forms of mobility. It then highlights the importance of *cosmopolitan capital* as a defining feature of elite cosmopolitanism. The discussion also actualizes an ambiguous mode of spatial appropriation that oscillates between cultural-material mastery and surrender (see Hannerz, 1992). In the following section I take a closer look at the field of international politics, development and diplomacy – understood as an order of recognition – to assess what it takes to become an elite cosmopolitan within this field. My analyses of the above-mentioned material show how the primacy of cosmopolitan capital gives a particular shape to the communicational doxa of the UN system. The subsequent two sections focus on the ways in which agents within the field manage the opportunities and demands associated with expanding orders of transmedia technology. I show how the field normalizes certain modes of communication and media use and, conversely, how broader patterns of everyday media dependence influence the field. A key finding is that the normalization of new media occurs largely indirectly, through non-formalized processes, and ultimately changes the conditions for acquiring cosmopolitan capital. At the same time, however, the basic logics of the field remain intact, implying that the appropriation and mastery of various media resources does not have any symbolic value as such.

Mobility, place and cosmopolitan self-identities

The word "cosmopolitanism" tends to evoke mixed feelings. Sometimes it is associated with visionary ideals of global openness, hospitality and mutual recognition between different social and cultural groups. At other times it is used as a label of culturally and economically privileged groups who can engage in frictionless international mobility. In the latter case, the labelling of certain persons as cosmopolitans opens a gap between "us" and "them" (the cosmopolitans), ironically evoking feelings of exclusion rather than inclusion. The reason for this division is that cosmopolitan values are socially stratified. Cosmopolitanism is empirically associated with higher levels of education and relatively mobile and urban lifestyles (see, e.g.,

Pichler, 2008; Mau et al., 2008; Lindell, 2014). It means that there are privileged groups in society (at least in the European context) that are good at *expressing* cosmopolitan values (in terms of hospitality, cultural openness, and so forth) and thus classify certain associated skills and lifestyle practices as distinctively cosmopolitan (see also Thurlow and Jaworski, 2006; Elliot, 2014).

When studying cosmopolitanism, therefore, we must acknowledge two things. First, as stated in Chapter 4, cosmopolitanism is ultimately to be understood as an *ethos*, an ethically grounded outlook and practical orientation to the world that includes self-reflexivity and hospitality as key virtues (e.g., Dikeç, 2002). Second, the cosmopolitan ethos is *culturally classified* (e.g., Bourdieu (1979/1984), meaning that the socially positioned and classified agents who in various ways express cosmopolitan values locate these values in social space as part of their classifying recognition work. It also means that certain properties and practices that are *not* to be understood as cosmopolitan from an ethical point of view, related for instance to gentrifying urban lifestyles and leisure cultures (see Chapter 7), may still evoke such connotations because of their association with other "cosmopolitan" forms of (self-)representation. These processes of classification thus (re)produce the gap between inclusive and exclusive connotations of cosmopolitanism.

In addressing this issue I want to advance a Bourdieusian understanding of cosmopolitan capital as a *sub-form of cultural capital*. The Bourdieusian approach (see, e.g., Bourdieu, 1972/1977, 1983) implies that there is a particular social field in which this form of capital is at stake and in which particular institutions of legitimation ensure the value of certain properties and skills. The usefulness of such an approach is that it allows for a sociological and critical understanding of how cosmopolitanism comes to surface within some realms of society as a desirable outlook associated with certain lifestyles without dismissing cosmopolitanism altogether as an elitist project, or, on the other hand, treating all expressions of cosmopolitan capital as grounded in ethically "pure" forms of cosmopolitanism. In other words, the Bourdieusian framework and the notion of cosmopolitan capital can help us turn the cosmopolitan paradox into a researchable area of symbolic struggle.

The meaning of cosmopolitan capital has been discussed in a number of recent texts (see Kennedy, 2009; Bühlmann et al., 2013; Igarashi and Saito, 2014). While there are variations among these analyses as to how cosmopolitanism is integrated with Bourdieu's theory of social fields, there are also common denominators. Above all, cosmopolitan capital is seen as entailing certain *embodied dispositions and competences* that reproduce cosmopolitanism as an ethical outlook. Key elements of cosmopolitan capital are language skills and the ability to deal with cultural differences in an open and sociable manner (see, e.g., Hannerz, 1990; Delanty, 2009: Ch. 2; Glick-Schiller et al., 2011). Furthermore, there are certain *institutions of legitimation* that recognize and reproduce the value of cosmopolitan dispositions and competences within a given social field. Such institutions are typically found within the education system and among international organizations and firms. Previous research has shown that cosmopolitan capital, especially in the objectified form of international degrees and career paths, has become increasingly valuable within the corporate sector (e.g., Carroll, 2009; Bühlmann et al., 2013). Nonetheless, the epicentre of cosmopolitan capital can be found in those sectors of society where cosmopolitan dispositions and skills are recognized as the key resource and measure of success rather than subordinate to other forms of capital.

Altogether, this means that cosmopolitan capital is closely associated with certain ways of life – in terms of professional trajectories and lifestyles at large – and certain forms of self-identity, rather than others. Above all, the classical elements of international mobility and regular encounters with foreign people, places and cultures, whose importance was discussed by Kant (1795/2003) as early as the eighteenth century, still seem to be essential to cosmopolitanism in times of extended mediatization. Those who can benefit from cosmopolitanism; that is, those who can take on a privileged form of cosmopolitan self-identity – which is different from other, more "ordinary" or working-class based articulations of cosmopolitanism that have been discussed in the literature (see, e.g., Werbner, 1999; Lamont and Aksartova, 2002; Nava, 2007) – have the capacity to engage with the Other, but also to stay in control of these encounters. As Hannerz (1992) points out in a classic essay, the privileged cosmopolitan builds his/her identity on a peculiar interplay between cultural mastery and surrender.

> Competence with regard to alien cultures for the cosmopolitan entails a sense of mastery. His understandings have expanded, a little more of the world is under control. Yet there is a curious, apparently paradoxical interplay between mastery and surrender here. […] The cosmopolitan's surrender to the alien culture implies personal autonomy vis-à-vis the culture where he originated. He has his obvious competence with regard to it, but he can choose to disengage from it.
>
> *Hannerz, 1992: 253*

Cosmopolitan identity thus involves a special sense of autonomy, the autonomy to engage with or disengage from foreign as well as familiar cultures. In order to acquire the particular competences that cosmopolitan trajectories require, the individual must learn about and engage with the Other and ultimately surrender to ways of life that are different from his/her own. However, inasmuch as cosmopolitanism can be recognized as a marker of social status, the individual cannot afford to get stuck in any one place but should actively manifest his/her sense of mastery and independence vis-à-vis *different* cultural contexts. Elite cosmopolitans are thus defined by an intricate balance between place-based immersion and continuous mobility, between surrender and mastery. They always have the opportunity to move on, or move back, from one place to another if things get too problematic. They are therefore, at least to a certain extent, in control of their own mobility – a specific form of mastery, and a manifestation of power, that Kaufmann (2002) terms *motility*.

These cosmopolitan dynamics were clearly exposed in a small research project I carried out in 2008 among Scandinavian development workers, most of them holding mid-rank positions and leading expatriate lifestyles in Managua, Nicaragua (see Jansson, 2009, 2011; Christensen and Jansson, 2015: 39-42). These workers expressed, on the one hand, a strong desire to get involved with the local population and not become too isolated in expatriate enclaves so as to learn as much as possible about the life-conditions and social problems they were supposed to work with. On the other hand, those who were senior professionals, who in some cases had been globally mobile for decades, described how the sense of growing mastery that comes with continuous encounters with different types of societies can lead to a sense of fatigue, even disengagement. At a certain point, the type of work they carried out even *demanded* disengagement in order to be able to move on to the next job and the next posting.

Similar experiences were found among the Scandinavian UN professionals that I interviewed in Geneva in 2014. They often spoke warmly about the countries they had worked in, but also highlighted the demands of continuous mobility the UN system placed on them in order to prolong their careers. The autonomy of motility, and thus cosmopolitan capital, thus stood out as a *relative asset*, largely dependent upon the policies and material conditions of different organizations. Peter, a portfolio manager in his forties, formulated this ambiguity in a rather straightforward manner:

> I have worked in more than ten different countries since I came to this organization. Previously I worked in a lot of countries in Latin America and in another African country. Now I have worked in these two countries [Angola and Rwanda] for almost two years so I feel that I've developed quite deep connections with these countries... a bit like a love–hate relationship sometimes. There may be very big problems to solve in these countries, but at the same time you get, or at least I get... I feel really strongly for them. I really want there to be positive changes in the countries I'm working with. [...] The schools are very good here [in Geneva]. It's rather comfortable, good working conditions. But it doesn't mean we're going to stay here forever. On the contrary, I would be very happy to move. [...] As I said, I don't like the city of Geneva at all. I try to avoid it as much as possible. So we're living outside of town... You know, in Geneva and this region there are about 40–50 per cent foreign-born people and if they work in an international organization like we do, then you can end up in a little bubble. The friends we meet are also working in international organizations. There is not much interaction with the rest of Swiss society. I don't really know what the Swiss people are doing.

What is particularly interesting about Peter's reflections is that Geneva, a foreign place for him even if culturally rather similar to Scandinavia, is not embraced at all with the same kind of cosmopolitan curiosity and engagement that more transitory contexts – that is, countries and regions that are perceived as "the field" (as opposed to the locations of international headquarters) – tend to evoke. Rather, the international "bubble" of Geneva is here taken as a superficial construct that could have been located almost anywhere in the world, and what lies beyond this expat "bubble" remains uncharted terrain. There are of course UN expats who are more engaged in Swiss society than Peter is, especially those whose lifestyles lean more towards the cultural and creative sectors (see Chapter 7). Nevertheless, what remains an essential ingredient in the shaping of cosmopolitan identity is the *mixture* of different cultural experiences, among which experiences from places and cultures that are considered more "alien" and more "problematic", and thus involve more friction, attain a particular status.

The interplay between global mobility, foreign experiences and cosmopolitan self-identity has also been confirmed in statistical analyses. In a survey conducted in 2014 among Swedish citizens living abroad we found that expatriates agreed to label themselves "world citizens" to a greater extent than Swedes living in their home country (Jansson and Lindell, 2015). Only a small fraction of the expatriates, about 20 percent, stated that they strongly or partly disagreed with the statement that they could be called "world citizens". In contrast to national surveys, there were no significant differences between social groups based on their level of education or social position, though it should be noted that a great majority of the expatriate population consists of people with higher education and white-collar jobs. Rather, what turned out to be the key to cosmopolitan self-identity was *high levels of mobility*, both in terms of how many different countries the respondent had lived in since moving from Sweden and in terms of ongoing mobility; that is, regular international travel to countries other than Sweden. There was also a significantly larger fraction among Swedish expats living in geographically remote parts of the world who considered themselves to be world citizens than there was among those within Nordic and European regions.

Cosmopolitan identity thus rests upon *multi-local engagements* (see Gustafsson, 2009). It requires continuous global mobility as well as local investments that follow institutionally and/or socially legitimized formulas. Ultimately, cosmopolitan identity *needs the Other*, and needs to *recognize the Other as the Other*, in order to establish itself as cosmopolitan and gain recognition within its own circles. This leads to a contrary situation in which the cultural privilege of elite cosmopolitans, regardless of their level of ethical engagement in global issues and disadvantaged populations around the world, is pre-conditioned precisely by the non-privilege and lack of motility among nominated Others. Accordingly, as Calhoun (2002: 893) reminds us, we must never fail to recognize "the extent to which cosmopolitan appreciation of global diversity is based on privileges of wealth and perhaps especially citizenship in certain states". This brings us back to the main point of this section: that is, the accumulation of cosmopolitan capital does not stand in a direct relationship with the development of a cosmopolitan ethos but is just as much associated with the reflexive habitus and recognition work of mobile middle-class fractions (Chapter 4). In the following section I discuss these dynamics in relation to Bourdieusian field theory.

Understanding the trajectories of elite cosmopolitans

> The limits of the field are situated at the point where the effects of the field cease.
>
> *Bourdieu, in Bourdieu and Wacquant, 1992: 100*

In order to understand the role of mediatization within the mobile world of UN professionals, we must first of all describe the logics of the field. This means that we need to specify what resources and symbolic markers are at stake among agents who want to gain power and status within the field (see Bourdieu, 1983). It should be stated here that the current material does not allow for any exact descriptions of the boundaries or dynamics of the field, especially not in a statistical sense. Still, the working hypothesis that will be sustained throughout this analysis is that the system of UN organizations constitutes the epicentre of a field of international politics, development and diplomacy where the formalized goal is to achieve, or produce, cosmopolitan forms of social change and where agents to a great extent operate within a shared job market.

As shown in previous research (e.g., Eriksson Baaz, 2005; Jansson, 2009, 2011) it is expected of international professionals within these sectors that they move between different organizations and different geographical postings during their careers. Many UN organizations maintain formal mobility and/or rotation policies, meaning that employees are either forced or encouraged to move to different locations on a regular basis in order to achieve a fully fledged understanding of the global field of operations. These specific conditions also generate distinctive social trajectories, understood as "constructed biographies" (Bourdieu, 1983: 346n), in which the accumulation of capital is closely tied to particular modes and sequences of global mobility. One could say therefore that the UN system *produces* elite cosmopolitans. The formal mobility policies and ethical principles that distinguish the UN system imply that UN professionals are required to accumulate cosmopolitan capital in order to reach more prestigious positions within the field. Agents in this field thus become elite cosmopolitans almost by default. The flipside of this distinction is a peculiar kind of precarity; besides the fact that agents may feel that they are forced to move on a regular basis, they may also find it difficult to translate their international experiences into equally qualified jobs in their countries of origin, should they wish to return. As one of the interviewees in the UN study puts it:

> I think there is age discrimination in Sweden. I'm not sure if it's about to change but I've read horrifying articles in Swedish newspapers about 41 year-olds being wasted in the job market. I don't think it's like that here [in Geneva]. [...] Especially in communications it seems like experience is less important than having 5,000 followers on Twitter. That's how it feels... And there are very few jobs [in Sweden] where international experience is important.
>
> *Matilda – information officer in her forties*

Although this type of social risk is a necessary part of the career trajectory, there seems to be a mutually confirmative relationship between the habitus of professionals within the UN system and the logic of the field. Only one informant stated that she had ended up in this type of organization by coincidence (head-hunted because of her managerial experience). All the others pointed to deliberate, long-term educational investments aimed at reaching this type of international position, combined with outspoken ambitions to "make a difference in the world". Their ethos thus coincides both with the cosmopolitan objectives of the UN system and the mobility policies that distinguish these organizations. There are some variations between individual organizations, however. Whereas the UN High Commissioner for Refugees applies a strict rotation policy, for instance, obliging its employees to move between countries to different offices and headquarters every second year, the International Labour Organization maintains more flexible regulations. Nonetheless, mobility in general and international field experiences in particular remain vital pre-conditions for ascending trajectories across the field.

> There doesn't have to be a written policy as such, but there can be internal pressure to also work in the field. One gets greater legitimacy if… yes, in one's profession if one has been out in the field. If one works with policy issues, for example, it's quite important.
>
> *Tina – technical officer in her thirties*

The interviews also reveal that the question of legitimacy becomes particularly obvious in cases where higher managers without international field experience are recruited to the organization. These managers have to work harder in order to gain respect. Another consequence of more open-ended mobility policies is that personnel sometimes resist moving on after they have reached Geneva headquarters, anxious to avoid getting stuck in a less comfortable place; that is, in a developing country where family life is more complicated. In organizations that apply formal rotation policies, by contrast, there is no choice but to move, and what counts is the type of postings the agent takes up during his/her professional trajectory. In such contexts field experience becomes even more important and, as one informant put it, "the tougher the postings one can get, the better".

Not surprisingly, the accumulation of cosmopolitan capital is difficult to combine with family life, especially if there is to be equality between partners. In certain places and on certain missions it is not even possible to bring accompanying family members. The women in the sample were either singles, single parents or living with a man who was willing to take the main responsibility for the family rather than having his own career. Dorthe is in her thirties and works for a UN organization in Geneva as a social protection officer after having spent six years in different transitional countries. She was previously in a relationship but found it difficult to make it work, given the demands of her professional life. Her story reveals that the rotation policy imposes a certain rhythm for making life decisions and coping with social and existential ambiguities. Ultimately it becomes a question of whether to accept the rules of the game or choose another career.

Dorthe: Right now it's a period when I don't know exactly what will happen, if I'm going home [to Denmark] or if I will continue, or what will happen. […]

Interviewer: How do you feel about the rotation policy?

Dorthe: As an organization, I can understand why they do it. It's a good thing, it's so easy to sit in headquarters and forget what we are actually working for, so in that respect it really makes sense. […] But in relation to other parts of life it's really a challenge. You need to have a rather flexible family life. Not everybody has that. And it creates a certain imbalance between men and women.

Interviewer: Is it more difficult to be a woman in this situation? What is your experience?

Dorthe: Mmm… I think, at least, if you look at our statistics it's easy to see that there are more women who don't have children in [this organization] and there are more women who are divorced, and it may have something to do with the culture of [this organization] but also with the world we live in and other structural factors. And in some places where we work we cannot bring our family, so there are many people who travel back and forth.

The UN system thus generates global mobility not only when agents move from one job to another, but also in terms of the logistics needed to keep families together. Similar experiences can be identified across the sample, even though life in Geneva is described by most informants as relatively easy and manageable compared to conditions they have experienced elsewhere. The material imperatives of the field may vary to some degree as agents move from one geographical location to another, but there is still an overarching pressure to be ready to move and willing to make certain sacrifices in order to maintain their positions within the field.

This uncertainty affects not only family life but also social relations in broader respects. Tina is committed to pursuing her career within the UN sector, which probably means moving to a foreign office after her current period in Geneva. But she is also aware of the risks of losing touch with some of her friends.

Tina: I have a very bad conscious… for example, in relation to old friends in Finland, when I say that I cannot go to Finland. But I cannot make any plans. Same thing in Indonesia, I couldn't plan anything. I sometimes had to cancel my vacations. So what happens is that I never buy my tickets before I really know that "OK, I'm going". Planning in advance doesn't work. Those who have similar jobs, we understand each other and don't take it the wrong way, but...

Interviewer: It's almost like a stand-by mentality.

Tina: Yes, that's what it becomes.

Interviewer:	You must be prepared to…
Tina:	But it's easier here in Geneva, because I don't have that many trips.

For many interviewees, especially women with young children, the Geneva posting seems to provide an almost parenthetical time-space for negotiating and reflecting upon their trajectories. Some of them describe how they actively try to loosen up the institutional pressure on them to be in a state of readiness, something that would have been unthinkable if located in a foreign office. Linn, a technical officer who is also engaged in union work and issues related to gender equality within her organization, mentions that she sometimes tries to say no to work-related trips, suggesting alternative solutions. However, even though she wants there to be more room in her organization to say "no", she is not willing to cross the line and jeopardize her future prospects.

Interviewer:	Does it affect your career to some extent, your level of mobility?
Linn:	Well, it does… Or our culture says that it does, but then I don't know if that's the case in reality. I would also like to think that it earns you respect, to "lead by example", kind of… […] There is a vague organizational pressure that it's not considered good to go home at 5.00 pm and it's not good to confess that one has a family and that they have needs. One should put the organization first. […] My immediate boss is quite sympathetic when it comes to my family situation and accepts that I sometimes say no − *but* there is always this undercurrent that "I actually want you to go on this business trip after all". So it's a difficult balance, which may influence your qualifications or stamp you as being uncooperative.

Linn's experience allows us to question the gendered value of cosmopolitan capital, exposing the limits of what goes and what does not. The field of UN organizations is marked not only by cosmopolitan values of global mobility and continuous engagement with the Other, but also by patriarchal norms that pre-suppose full commitment to the organization at the expense of family life. Access to cosmopolitan capital is thus conditional on broader structural factors such as gender and family situation, which means that women who want to reach elite positions must either develop strategies for sidestepping some of the taken-for-granted expectations regarding continuous mobility and flexibility, or organize their family lives in non-patriarchal ways. In this context, as we will see, mediatization plays an increasingly important role, both as an emancipatory force (providing greater autonomy to travel) and as an extension of the social field (including extended forms of media dependence).

The indirect force of mediatization

While mobility can be seen as a direct source of capital accumulation among UN employees, media practices have a more indirect significance. The interviews reveal that obedience to communicational doxa does not have any value in itself, but is subordinate to the demands to be mobile and acquire international competence. Mediatization thus constitutes an indirect force in this regard (see Hjarvard, 2008: 114–15). There are certain informal rules and expectations about which media are considered mandatory and to what extent individuals are supposed to use them, but they are relatively volatile across organizations and depend on individual leadership and geographical location. Again, there is a general consensus among the informants that Geneva provides an "easier" environment than foreign offices where there is little room for social strategizing.

Leena (programme officer):	It's more OK here than in Zambia. When I was there it was too much. Most often I had to look at the subject line to see if an email was something I had to read immediately or not and if I could delegate. I had too much. I couldn't read all my e-mails.
Interviewer:	Why was that? What was it that…?
Leena:	I had a rather central role, I was directly below the boss and Country Director. Everything, or almost everything he got, I suppose, I got too. Or if he delegated to me, then I had to find somebody who could do it or do it myself. Yes, it was because of that central role everything was copied to me. And then of course I was also expected to read all my e-mails and respond to all my e-mails within a certain time. It was a lot of different things. It could be budget issues, it could be political issues, it could be all kinds of things...

E-mail is regularly mentioned as the prevailing means of professional communication. Informants who occupy intermediary or higher positions describe a situation in which information overload often occurs because of the routinized habit of copying a number of staff members into e-mail correspondence. There are no written policies behind these norms; they stem rather from the desire to make sure that nobody feels sidelined within a project. This, in turn, is related to the general importance of making one's work visible, which in the long run may have a positive impact on the agent's chances of advancing within the field.

Interviewer:	I was thinking about one thing… What is it that counts the most in your professional life? What do you have to do if you want to move on in your career from where you are now?
Leena:	One has to be visible in some way, in a good way. It's not enough to do one's job. One also has to do it well. And then people need to know about that work one has done…
Interviewer:	And how do you do that?
Leena:	For example, in the project in Spain a great deal was visible from the start, same thing with Greece. The project itself gained a lot of

	interest. If you do a good job people will notice. But you also have to know the right people. That isn't so easy.
Interviewer:	And then you have to attend the right events or what…?
Leena:	As I said, I cannot say no to business trips. In that regard I cannot say no.

As reported from other expatriate environments as well (see, e.g., Beaverstock, 2002; Jansson, 2011; Polson, 2016), the ability to connect with the right people (both online and offline) and to make one's work visible, acquiring what Urry (2007) calls "network capital", is important for not getting stuck in one's career. Online networking platforms like LinkedIn and Facebook are seen as more or less indispensable by the informants. The use of these platforms has emerged and expanded gradually without any formal sanctions and, it seems, without much reflection by the agents themselves. However, while they are treated as parts of the taken-for-granted lifeworld, spanning the borders of private and professional life, their mastery becomes valuable only in as much as it contributes to the accumulation of cosmopolitan capital.

These findings seem to diverge from those of my earlier studies conducted in 2008 among expatriates within the field of international development aid (Jansson, 2009, 2011). In the latter case, attitudes towards new networking tools and the circulation of personal information were marked by a great deal of reflexivity, even scepticism. Cultural (cosmopolitan) capital was found to work as a resistant force. The most important reason for the divergence between these studies is probably the time factor. While social media platforms were not yet naturalized in 2008, most agents within the field have since naturalized certain ways of coping with the pluralized media landscape. Even though the organizations do not make any explicit demands vis-à-vis networking, or even provide all their employees (but only those in higher positions) with such hardware as smartphones and tablets, agents generally consent to the communicational doxa, thus reproducing the overarching logic of the field. The story of Matilda that introduced this chapter provides a fairly typical illustration in this regard. The fact that Matilda is deeply engaged in the kinds of issues she works with and is also eager to perform well at work leads to a situation in which private media as well as one's private home life are gradually, and without any resistance, absorbed by communicational doxa. She somewhat sardonically refers to this development, in which a transactional form of media dependence is normalized, as a "good-girl-thing".

Complicity with, and *reinforcement of*, implicit professional expectations are certainly not unique to the field of international politics, development and diplomacy, but are, rather, a foundational feature of doxa as such, according to the Bourdieusian understanding. However, as the present analysis has shown, the combination of (1) *emotional engagement* in one's professional area of interest, (2) *mobile working conditions*, sometimes involving coordination across time zones, and (3) access to *transgressive forms of digital media* leads to a steadily growing overlap between professional and private time-spaces, where media dependences easily spill over from one realm to the other. The mediatization of communicational doxa, in turn, implies that agents within the field are making investments in the field most of the time *without* altering the overarching principles of doxa. This illuminates how mediatization processes within this field – as well as within other fields, but in different ways (see, e.g., Kantola, 2014) – attain at the same time a socially transformative and structurally reproductive power.

There are of course variations depending on the organization and the career level of different agents. In general, however, my fieldwork suggests that the culture of UN organizations is marked by great adherence to the doxa of more or less permanent mobility. This concerns both the demands of work-related travel that are associated with many positions, and the expectation that agents will build professional biographies that include a certain mix of postings around the world. Along with these demands, as we have seen, come certain functional requirements regarding the use of media technics. In other words, the rules of the game are such that disobedience to communicational doxa and the questioning of certain means of communication always involve a risk, even though the possession and mastery of private media devices and platforms does not count as capital *per se*.

Media dependence and the osmotic expansion of doxa

One of the most remarkable things about contemporary mediatization processes is the largely informal, diffuse and contradictory ways in which they take shape. Digital transmedia technologies occupy a ubiquitous yet rather mysterious position in the lives of my informants. In certain respects, such media even belong to a "technological unconscious" (Amin and Thrift, 2007), especially when it comes to the seemingly natural ways in which they have, as a matter of routine, started crisscrossing the lines between private and professional realms. For instance, many of the directors, experts and technical officers I have interviewed are used to an almost overwhelming amount of job-related e-mails that cannot always be handled during regular working hours. These are combined with implicit organizational expectations of their availability, including when formally off duty. As mentioned above, they also describe an organizational culture in which it is common practice to copy a large number of colleagues into email conversations in order to ensure that certain measures are being documented and seen by the right persons and to avoid the risk that anybody feels sidelined. Linn, the technical officer quoted above, explains:

> If there is a report I have to submit then I also have to copy to x, y and z so that I'm covered, kind of, so there isn't anybody responsible for an area you were not aware of who becomes upset or sends an angry email back to you with a copy to the highest director... So it's used for marking territories, of course. And if you look at the level of the departmental director, he or she probably receives two hundred emails every day – enormous amounts... and it's very difficult to change even though you might try.

These types of normalized and routine procedures, established during the era of desk-bound computers, are part of communicational doxa and are today reinforced by the spread of personal media technologies. Smartphones in particular make it possible to handle email flows when on the move or at home. In the organizations I have visited, however, only managers at higher levels are entitled to receive such technologies from their employer, which means that doxa has successively come to involve an implicit material demand on agents to purchase and use personal technologies to carry out work-related communication. As Leena, the technical officer in her early thirties puts it, "When I worked in Zambia I had a Blackberry but not here. Don't ask me why […] but I'm expected to be reachable via my phone, which is *my* phone".

At the same time the pressures to be connected, visible and available often emanate from socio-technological changes that have taken place beyond the organizational framework. The social normalization of connective platforms like Facebook, Skype, Viber and WhatsApp, which are used regularly by most of the informants, is interwoven with the need to stay emotionally connected with family and friends. Such media offer a liberating potential that makes it easier for mobile households to organize their lives and compensate for the emotional and social costs of mobility, achieving a state of "connected presence" (Licoppe, 2004). This is particularly salient in situations like the above-mentioned challenges that women face in relation to the patriarchal doxa of the UN system. At the same time, however, these developments necessitate a composite and growing communication environment that generates dependences across the domains of everyday life, regardless of geographical location and time of the day or week. The normalization of individualized media access implies that family life may be further distributed across geographical spaces, as suggested by Christensen (2009), while at the same time reinforcing and extending the realm of doxa.

This *osmotic situation* is not unique to the UN, of course (see, e.g., Fast and Lindell, 2016). However, there are certain conditions that make the boundaries of professional life more permeable in this field than in most other parts of society. Most importantly, the mobile nature of these professions and social trajectories brings with it a heightened transactional and ritual dependence on various means of interpersonal communication (see Chapter 3). As Elliot and Urry (2010) note, digital devices come to serve as a kind of mobile depository of emotion, lowering the social and emotional costs of travelling lifestyles. Furthermore, the reduction in the economic costs of staying in touch with friends and family in one's homeland and elsewhere – a key aspect of polymedia (see Madianou and Miller, 2012) – has led to a radically broadened scope of media choice and more flexible interaction rituals. Many of my informants point to this lowered techno-economic threshold as the single most important change in their working conditions during the last decade. Peter, a portfolio manager in his forties, describes how he now uses his commuting time in Geneva to contact friends in other parts of the world:

> One important change is the costs and the barriers that there used to be for making phone calls from a mobile. The more those costs are going down the lower the barriers become, and it's much easier for me to keep in touch. It's always easier to call from the mobile than waiting until you get home

> in order to make a Skype call or call someone from a landline connection. […] When I drive home from work, that's when I call my friends most of time, in Sweden or wherever… That's when I feel that I have that time and I'm not doing something else or there are other things going on just around me. When I'm at home it's more difficult. […] In the car, I just give a call to people I come to think of in the moment. That's how it works.

This example highlights a sense of growing individual autonomy, the possibility to establish a genuinely glocal communicative space in the interstices of everyday textures where media practices amalgamate with other routines. Other informants tell similar stories, pinpointing a number of different channels and applications. Notably, instant messaging systems such as WhatsApp have, in some cases, been introduced by the interviewees' children and then appropriated as increasingly indispensable technics for communication and coordination among family members as well as close friends.

This is also the point, however, at which potentially greater autonomy runs the risk of abolishing the barriers to work-related communication, leading to the appropriation of applications like WhatsApp by communicational doxa. While email and Skype are the predominant means of communication within the UN doxa, WhatsApp and other social media provide opportunities for extending the reach of doxa and keeping agents in their place. Leena describes a situation in which her boss used WhatsApp to overcome a spatio-temporal gap in communication:

> So he wrote to me on WhatsApp during lunch, which is actually leisure time. "Where are you and when are you coming back?" and I answered that "I'll be back in ten minutes" and then went directly to him and wrote that "I'm here now"… But I hadn't read – he had sent me an email during the lunchbreak, which I hadn't read – and I just went straight to him and said, "I'm here now, what do you want to talk about?" and he said, "Ah, didn't you read my email?"

The important thing to note here is that there are no formal discussions or decisions taking place as to what media access should be expected from colleagues, or in what ways certain media should be used. Rather, new media are entering the realm of communicational doxa through processes of social transaction, supported by other unspoken dictums of the UN doxa, notably the requirement to be flexible, available and ready to move. In effect, we may speak of mutual *osmotic pressures* between doxa and everyday processes of mediatization, leading to the erosion of emotional as well as geo-social boundaries between work and leisure.

However, there is also resistance to the ways in which media influence doxa. As we saw in the very opening example of this book, there are UN employees who try to decrease their amount of travelling as much as possible, even try to avoid positions that would involve too much travelling, because they think digital media make professional mobility too stressful (see Chapter 1). Career adjustments such as these, which may lead ultimately to social stagnation, can also be seen as expressions of transactional media dependence, but in a negative sense. They illustrate a way of *not* submitting to communicational doxa; that is, of gaining a higher degree of autonomy as an individual subject at the expense of losing some of their autonomy as an agent within the field. For those who are new to the field and who try to combine family life with a cosmopolitan career, such considerations may be particularly critical, as heterodoxic agency, as we saw above, can have a negative effect on their trajectories in the field. While the limits of the field might be put to a test, it would be a difficult, not to say impossible, endeavour to negotiate the basic rules of the game. Continual mobility and the gaining of international experience remain the keys to capital accumulation. Media-related skills and resources, in turn, are to be understood as normalized material and cultural undercurrents that shape one's chances of gaining recognition within the field.

Conclusion: Mediatization as a problematization of privilege

In this chapter I have assessed the ways in which contemporary mediatization processes influence the logics of cosmopolitan fields, exposing how the emergence of various forms of media dependence plays into the creation of cosmopolitan elite identities and trajectories. The analyses started out from a critical understanding of mediatization as marked by the dialectical interplay between autonomy and dependence, and applied qualitative interview data gathered from Scandinavian expatriates working for international organizations in Geneva to gain an inside view of the ongoing adaptations to changing orders of technology.

The conclusions can be summarized in three main points. First, the interviews revealed that mediatization processes exercise merely an *indirect influence* on the logic of the field. While the general conditions for acquiring cosmopolitan capital, understood as a field-specific sub-form of cultural capital (Igarashi and Saito, 2014), have clearly changed along with the successive recomposition of communicational doxa (including the expanding taken-for-grantedness of personal devices for online communication), this does not involve any substantial shift in terms of what counts as capital (chiefly international experience and connections and a manifest readiness to be mobile). At the level of day-to-day activities, then, mediatization can be conceived as a complex set of environmental changes that accentuate rather than challenge the classificatory structures of the cosmopolitan field.

This underscores, secondly, that the dependences that mediatization brings about are established through the interplay between socio-cultural pre-conditions (orders of recognition) and technological affordances (orders of technology). It means that mediatization takes on different shapes in different settings (e.g., Couldry and Hepp, 2013; Kantola, 2014; Hepp et al., 2015; Jansson, 2015a) and that there may also be intertwined, sometimes competing, orders of technology and recognition. The transnational field of international politics, development and diplomacy, taken as an order of recognition, is marked by a situation in which new media habits, notably the uses of social transmedia platforms, are partly grounded in the socio-emotional needs of highly mobile and distributed family lives. The ubiquitous nature of transmedia technologies, in turn, makes them highly absorbable by communicational doxa, contributing to the geo-social expansion of the field, which then acts back upon the private lifeworld and its order of recognition. I have described this situation in terms of *osmotic pressures* that force individual agents to establish a sense of equilibrium between doxic demands and more existential needs. Similar processes have been identified in previous research on mediatized work/life-conditions (see especially Gregg, 2011), and in this particular field they gain accentuated prominence due to the doxic demand on agents to be mobile in order to achieve recognition.

Finally, the Geneva study has shown that the interplay between mediatization and the creation of cosmopolitan elite identities is conditioned by *place and gender*. Even though the overarching logics of the field persist regardless of location, the interviews revealed that the possibilities for tactical negotiations in relation to communicational doxa and other organizational demands are greater in Geneva than in foreign offices. This situation, in combination with the flexibility of new media, proved to be particularly important for female interviewees in the parental phase, as their capacity to make investments in the field is already circumscribed by the patriarchal structure of the UN system. The Geneva location was experienced as a parenthetical time-space where cosmopolitan capital could be consecrated for relatively low emotional costs in terms of broken ties and longer periods of absence from one's family. In a longer perspective, however, further investments in the field would necessarily require intensified mobility and thus greater pressures on these agents to act in accordance with communicational doxa. The contradictory experiences of mediatization, the tension between autonomy and dependence, thus went hand in hand with a feeling of being located at the existential crossroads between an extended cosmopolitan trajectory and return migration, where the latter would probably be the same thing as leaving the field.

These points lead to *a problematization of the notion of elite cosmopolitanism as a privileged structure of feeling*. While the individuals figuring in this study are clearly privileged in the sense that their life trajectories adhere to the ideal of self-realization, they also find themselves in somewhat precarious positions due to the fact that they either have to surrender to doxa or reconsider their mobile lifestyles and cosmopolitan self-identities altogether. Mediatization constitutes a regime that accentuates this tension. While mediatization is integral to the realization of emancipatory biographies, especially seen from a gender perspective, it also underpins and exposes precarity. My studies have shown how the appropriation of new media is rarely an independent choice but is integral to communicational doxa. Yet, within the UN system they are not commonly supported by institutionalized resources and not given any direct recognition, but merely so indirectly through the appreciation of networks that can spur further mobility and the accumulation of cosmopolitan capital. The ultimate question then is: *What do we mean by autonomy?* As discussed in Chapter 4, recognition constitutes a pre-condition for autonomy. However, as we have seen in this analysis, the gaining of certain forms of recognition, such as within a long-established social field, may also undermine the sense of individual autonomy in a broader sense. A key marker of elite cosmopolitanism as a structure of feeling is that individuals are basically engaged in a voluntary and relatively affluent form of mobility, but cannot control their mobilities on the more practical level – at least not so long as they want to remain within the field. Furthermore, when the demand to be mobile, and thus connected, increases, the regulation of osmotic pressures (the opening and closing of boundaries) becomes an escalating challenge. These dialectical conditions open up important questions of social power that I will return to in the final chapter of this book.

6

MEDIATIZATION AND POST-TOURISM

> Arriving at such a place knowing that you are the first person to enter in thirty-five years… It's like parachuting, which I've never done, but something really exciting. These are my parachute jumps.
>
> *Sven – urban explorer in his mid-fifties*

While adventurous leisure activities may seem emblematic of contemporary "re-enchanted" consumer cultures (see Ritzer, 1999), they have played an important role throughout modernity. In his essay "The Adventure" of 1911, Georg Simmel (1911/1997) discusses the significance of adventurous activities in early industrial society. Adventures come in many guises, Simmel argues, ranging from love affairs to mountaineering expeditions. What they have in common is that they take place outside the ordinary continuity of life. They constitute exceptional, sometimes spectacular, events whose outcomes are exposed to a certain amount of risk and thus remain relatively open-ended. At the same time they are anchored in the ethos of the adventurous person and are "distinct from all that is accidental and alien, merely touching on life's outer shell" (ibid.: 222). Adventures are deliberately created by the individual who is driven by a desire to experience something that is inherently authentic, non-replicable and uncoupled from the orders of everyday life.

The opening statement, taken from a longer interview with Sven, a middle-aged urban explorer, expresses this tension between careful planning and uncertainty, what can be described as "calculated risk". Sven works as an economist, but his main interest is to find, enter (often without permission), explore and photographically document derelict industrial spaces and ruins. While he maintains a deep interest in industrial and social history he is also fascinated by the spatial and aesthetic qualities – textures, vistas and volumes – of these locations, as well as the extraordinary experiences they evoke. It is thus no coincidence that he compares his unusual hobby to parachuting. Urban exploration – also called urbex or just UE – can be seen as a well prepared vertical jump into the time-spaces of bygone eras, whose decaying and ghostly appearances may elicit physically thrilling and uncanny sensations in their visitors. As Tim Edensor argues, industrial ruins hold the potential to confront the explorer with otherwise unheard stories of modern society:

> This impressionistic, spectral intimation of working life is radically different from the memories of industrial life expertly captured in museums and heritage centres, in which people are encoded and contextualized, categorized and narrated. Accordingly, ruins are places from which other memories can be articulated, from which "the things and the people who are primarily unseen and banished to the periphery of our social graciousness" (Gordon, 1997: 196) may be encountered. Such marginal locations are the most densely haunted spaces of the city, and following these ghosts allows us to pay our dues to the unheralded, spectral denizens who made the wealth of the city.
>
> *Edensor, 2005: 843*

Similar things can be said about other modern ruins, such as closed-down hospitals, hotels and military facilities that remain uncharted terrain to most citizens, located far off the beaten track, while at the same time evoking echoes of a strangely familiar past. Visiting such locations – as well as other sites of urban exploration, notably the hidden back regions of active buildings and infrastructures – can be seen as the logical contrast to touristic journeys and other staged experiences. It points to how the search for "pure" and "authentic" experiences gains accentuated cultural significance in times of technological rationalization and symbolic overflow. As Simmel (1911/1997) noted, while the adventure may look like an anachronism of modern life it should be understood as an integral part of it, precisely because it negates, and thus confirms, the dominant social order. Through its deviation from the norm the modern adventure epitomizes individual mastery and liberation from constraining forces, whether natural, social or technological. This also means that the modern adventure is closely related to the individual's self-identity. It is something more than just a mode of generating extraordinary sensations of body and mind. It becomes a classificatory practice that sets certain people apart from others.

So who are the urban explorers and why is this form of modern adventuring interesting in relation to mediatization? It does not take much investigation to conclude that urban exploration is an activity carried out mostly by men and that most urban explorers belong to the (broadly conceived) middle classes. One only has to browse through some of the many online forums and blogs that exist today to discover that urban exploration is not a *sub*-cultural activity in the stricter, class-based sense of the term, but rather constitutes a culturally distinctive activity that has arisen among relatively well-established and well-educated groups. This bias is also reported in previous studies (see, e.g., Bennett, 2011), even though it is difficult to find any precise demographic statistics. As Mott and Roberts (2014: 237) assert, the adventurous and self-realizing practices of urban exploration are not equally available to everyone, and the discourses surrounding it tend to "privilege particular bodies and spatial engagements, while discounting others". According to these same scholars, urban exploration can even be seen as a self-obsessed form of masculine play in which photographic practices create a false nostalgia that overlooks, even exploits, the life-conditions of those who inhabited the now-abandoned spaces (see also High and Lewis, 2007). The male dominance thus corresponds to deep-seated mythologies of Western masculinity that celebrate the penetration, conquest and colonization of foreign space. The ways in which many (male) urban explorers present themselves online seem to bear this out:

> Perhaps most typical is an ordinary global north metropolitan grungy look, though with deliberately selected functional footwear and accessories – especially cameras. No matter the attire, for many urbexers, recording themselves exploring appears to be very important and there is much emphasis on visual self-representation. Urbex websites and blogs are replete with images of urbexers posing on ladders, inside tunnels or abandoned corridors and so on. They are invariably in a conquering or heroic mode.
>
> *Mott and Roberts, 2014: 239*

Still, in spite of these hegemonic features, urban exploration must be acknowledged as a multi-faceted assemblage of practices that also entails culturally transgressive, even counter-hegemonic modes of appropriating and representing space. As noted above, Edensor (2005) sees the exploration of derelict places as a way of problematizing dominant forms of heritage production (see also Edensor, 2007; De Silvey and Edensor, 2012). Terms like "space hacking" (Dodge and Kitchin, 2006) and "place-hacking" (Garrett, 2010, 2012, 2014b) have been suggested as a way of articulating the activist ethos of urban exploration, seeing it as a critical reaction to the capitalist surveillance and commodification of urban space. Similar arguments have figured in public debates. For instance, in a commentary article in the *Evening Standard*, Will Self (2014) reacts to the prosecution of twelve British urban explorers who had been trespassing in various off-limit parts of the London transit system and had therefore, according to the transport authorities, "conspired to commit criminal damage". Self's point, however, is that urban explorers, seen as place-hackers, are "performing a valuable service by reminding us that the city should, in principle, belong to its citizens, and should mostly – if not entirely – be accessible to them".

It will be evident from this introduction, then, that urban exploration occupies an ambiguous social and cultural position. In addition to the general tension between calculation and risk-taking, between cultural reflexivity and inner sensations that characterizes most kinds of adventures, urban exploration gives us a lens through which we can investigate how middle-class mobilities interact with the dialectic of mediatization. Even though the raw, windswept experience of abandoned or derelict time–space locations may be at the core of urban exploration (see Garrett, 2014a; Hudson, 2014), the movement's growing popularity could hardly be imagined without media technologies. Above all, photography is essential to what we may call the *urbex communicational doxa*. Urban explorers are, more or less by definition, equipped with cameras and most often use individual websites and/or social media platforms to share their experiences with other explorers. When entering the search term "urban exploration" on Flickr, for example, one receives an overwhelming amount of hits, including an uncountable number of photos and hundreds of specialty groups dedicated to particular "sub-genres" of urban exploration – such as particular cities or regions, or particular types of sites (e.g., abandoned industries, military complexes or subterranean spaces). There are also mobile applications like *TalkUrbex* that enable members to interact and circulate their photos, as well as numerous websites, blogs and online forums. One can reasonably assume that the rapid expansion of the Internet in general and connective media in particular has played a significant role in the popularization of urban exploration.

Inasmuch as urban exploration is a visual and social practice it is thus also a media-dependent practice, helping to advance the cultural forms of connective media. However, *the ways* in which these media practices are played out, and how they relate to different forms of mobility and spatial appropriation, are far from clear-cut or uniform. Rather, they are marked by the dialectic of mediatization, which is in turn reinforced by the ambivalences of middle-class identities.

In this chapter I discuss urban exploration as a potentially counter-hegemonic force, an emerging structure of feeling that develops *within* mobile lives (understood as the overarching, hegemonic structure of feeling) as a reaction and alternative to dominant forms of mediatized mobility, notably mainstream tourism. However, as pointed out in previous analyses of so-called *post-tourism* (see, e.g., Feifer, 1985; Urry, 1990; Munt, 1994; Jansson, 2002a), the potentially subversive turn to "other" spaces, in this case half-forgotten or hidden places of modern life, should at the same time be seen as an expression of the cultural sensitivity and classificatory struggles of upwardly mobile middle-class groups. Their media practices, as we will see, are largely governed by an aestheticizing and intellectualizing gaze, whose self-reflexive qualities in the longer perspective may become constitutive of popularized (and thus less distinctive) urban imageries (e.g., "ruin porn") and hegemonic forms of spatial consumption (e.g., "ruin tourism") (see, e.g., Millington, 2013; Tegtmeyer, 2016). As in the context of gentrification (see Chapter 7), the reliance on symbolic circulation for manifesting uniqueness and individual discovery has a tendency to erode the authenticity it once sprang out of. Accordingly, many urban explorers (at least in the sample consulted for this study) were found to be strikingly reflexive as to how they depict and share information about the sites they visit.

The chapter is divided into five sections. In the first section I position the phenomenon of urban exploration within a broader cultural and social context. This means that I address the middle-class-biased and post-touristic nature of urban exploration. In the three subsequent parts of the analysis I look more closely into the spatial and communicational doxa of urban explorers, pointing especially to how the continuous interplay between socio-cultural reflexivity and the quest for existential authenticity shape "urbex media" as a cultural form. These sections of the chapter look at three separate (albeit inter-related) stages of spatial appropriation and representation: site discovery, camera work and storytelling. In the final part, by way of conclusion, I go back to Williams's original thoughts and assess the structural significance of urban exploration as an emerging structure of feeling and counter-hegemonic critique of mediatization.

Urban exploration as post-tourism

Urban exploration is a relatively new phenomenon, even though its roots stretch back at least to the nineteenth century and the exploration of subterranean urban spaces (such as catacombs and transit tunnels). The term was allegedly coined in 1996 by the Canadian urban exploration pioneer Jeff Chapman, also known by his alias Ninjalicious. In his book *Access all Areas: A User's Guide to the Art of Urban Exploration* Chapman defines urban exploration as "an interior tourism that allows the curious minded to discover a world of behind the scenes sights" (Ninjalicious, 2005: 3). The parallel with tourism is not coincidental. Indeed, one can hardly discuss the meaning of urban exploration without relating it to tourism. According to Wikipedia (information retrieved in April 2016), urban exploration is "the exploration of man-made structures, usually abandoned ruins or not usually seen components of the man-made environment. Photography and historical interest/documentation are heavily featured in the hobby and, although it may sometimes involve trespassing onto private property, this is not always the case". Urban explorers, therefore, are mobile people willing to travel long distances in their free time to discover, enter and experience new sites, appropriating them as adventurous time-spaces as well as visual worlds. They also document and share their experiences with other explorers, letting the rest of the community know that they have visited a certain site.

Urban exploration thus entails a form of spatial consumption and circuits of representation that in key respects resemble tourism (cf. Jansson, 2007a). There are interesting overlaps between urban exploration and several "sub-genres" of tourism, such as heritage tourism (an interest in historical sites and ruins), dark tourism (an interest in medical spaces of life and death) and eco-tourism (the willingness to uphold a strong ethical stance in relation to the sites visited) (see Robinson, 2015). But urban exploration is also distinguished by its *antagonistic* stance vis-à-vis tourism; it is actively defined in opposition to tourism and touristic practices. Robinson (ibid.: 149) holds that urban exploration can be viewed as "a form of touristic activity, and, paradoxically, anti-tourist resisting all the formal tourism practices of signage, information, instruction and control". The sites and sights of urban exploration are by definition located *beyond* the radar of commercial tourism. They are sought out precisely because they are *not* named, framed and mechanically or socially reproduced as tourist sights – referring here to MacCannell's (1976: 44–45) classic four-stage model of "sight sacralization". Whereas ruins are understood as accessible and marked-out places, preserved by public authorities for the sake of tourism and heritage, the types of places that urban explorers visit are not open to the public. This distinction was also expressed in several of the interviews with Swedish urban explorers.

It's a very broad expression and it feels like it's really lost focus lately, what people have called urban exploration… I wouldn't call it urban exploration to, like, join a guided tour at some old fortress with the family, that doesn't feel very much like urban exploration. That would be more like ruin tourism. […] If you compare urban exploration with ruin tourism, for example, in ruin tourism you just go to ruins that are open and accessible to everyone, like an old factory that nobody cares about. And if you would never consider climbing into a place then I think you are ruin tourist and not an urban explorer.

Sven – urban explorer in his mid-fifties

Urban exploration is thus a socio-cultural phenomenon that incorporates symbolic distinctions. Besides the adventurous desire for raw and unmediated experiences, urban explorers want to separate themselves from the tourist crowd. They are representatives of the anti- or post-touristic sentiment that can be found especially among mobile middle-class groups whose social status might benefit from the problematization of established symbolic hierarchies (see Chapter 4). The term post-tourism is in my view better suited than anti-tourism for unpacking what urban exploration is about. Following Munt's critique (1994), I will apply the term post-tourism in a sociological and relatively inclusive manner rather than seeing it as just a form of reflexive sign-play. From such a socio-cultural perspective, post-tourism not only illuminates the negative relationship between urban exploration and mainstream tourism, but also, and more importantly, locates the phenomenon in relation to transformations of social space in late- or postmodern societies. Although the key works on post-tourism were actually published *before* the coining of the term "urban exploration" in 1996, they refer to touristic phenomena that in many ways anticipate the urban exploration movement. In her original discussion of post-tourism, Feifer (1985) pointed to the ongoing, and potentially disruptive, *pluralization* of tourist practices associated with the expansion of the (new) middle classes. This includes the postmodern enjoyment of moving across pre-established taste patterns and seeking out semiotic effects by conducting tourism in unexpected ways, within as well as beyond dominant forms of mass tourism. Along similar lines, Urry (1990) described the emergence of specialized travel agencies responding to the desire among middle-class fractions to experience "real", that is, non-staged, destinations (see also MacCannell, 1976: Ch. 5) and travel in *flexible, individualized* ways (see also Lash and Urry, 1994). Other important features of post-tourism that have been noted include *de-differentiation* – that is, mixing tourism with activities that are not necessarily touristic, especially intellectual interests (related to anthropology, archaeology, etc.) and what I described above as adventures (mountain-climbing, trekking, etc.) – and a concern with Otherness, reflected in cosmopolitan desires to understand minority cultures and travel in culturally and environmentally sustainable ways (Munt, 1994).

Urban explorers express attitudes that are largely convergent with all of these features of post-tourism, even though their explorations rarely include culturally or globally remote places. Rather, they are *travellers in time* who actively engage with and recognize human lives and place biographies of the recent past. Their visits to abandoned places are framed by ethical codes that can be summarized in the well-known (and in the interviews continuously repeated) urbex dictum: "Take nothing but photos, leave nothing but footprints". In this way they also distinguish their community from other urban adventurers, especially graffiti painters:

> Their interest is to paint and some are really good while others just do tags and destroy things… It's not very often we run into one another, after all, because most of these industrial places we're looking for… it's like our paths never cross. I don't have any good answer but we don't belong to the same kind of group. They prefer to paint in tunnels and at railway stations and coaches, where it's visible. Most of them don't want to paint where it's not visible. But sure, if I look at my webpage I would guess that eighty per cent of all places I've been to have some kind of mark, like "I was here". […] You should take great care to leave the place exactly like it was. That is the worst thing I know, when people have been there and painted graffiti all over the walls. It really pisses me off. If I were to choose, everything would remain as it is for another thousand years… It's extremely interesting, and you hope that others will be able to experience it in exactly the same way.
>
> *Niklas – urban explorer in his mid-thirties*

Based on the above discussion I suggest that we conceive of urban exploration as a post-touristic expression and extension of the type of middle-class identity and ethos that gained social prominence from the 1970s onwards. As such, urban exploration incorporates social ambiguities that point ultimately to the emblematic forms of recognition work that saturate middle-class lifestyles.

There are two areas of tension in particular that define the orders of recognition through which the adventurous, post-touristic identities of urban explorers take shape (see Figure 6.1). First, there is a tension related to how the *relationship between self and site* is negotiated. On the one hand, as shown in the first interview extract, urban explorers think of their explorations as a search for a particular kind of authenticity, one that is "true" in relation to both site and self-identity. Their mode of appropriating decaying industrial landscapes can be described as neo-Romantic (see, e.g., Garcia, 2016) and corresponds to how Wang (1999) sees the role of *existential authenticity* in certain forms of tourism. In Wang's view, existential authenticity is different from objective and symbolic forms of authenticity because it is achieved only inasmuch as the individual actor senses that there is a meaningful relationship between the history of the site and the biography of the self. Existential authenticity thus means that the site speaks somehow directly to the visitor who is dragged into the unknown, yet strangely familiar spaces of the past (see also Steiner and Reisinger, 2005).

On the other hand, as shown in the last interview extract, urban exploration is marked by high degrees of *socio-cultural reflexivity*. While the search for existential authenticity denotes a desire to be immersed in space, there is a parallel movement towards (photographic) aestheticization and intellectualization of the historical meanings of sites and how these meanings have transformed over time (Munt, 1994). In the Swedish interview data there are several examples of urban explorers specializing in certain areas of aesthetic or intellectual expertise, such as industrial production methods, architecture or cultural history. This kind of expertise also grants the beholder a certain status within the community.

The following extract, also taken from the interview with Sven, shows how the tension between existential authenticity and socio-cultural reflexivity is played out. It also highlights how an ambiguous relationship to place resonates with a middle-class habitus marked by social mobility.

> This Cementa place, the limestone quarry, it has been the life support of several of my old classmates' fathers. The prosperity we have today was built on places like this. And today we sit and look at one another in these small cameras in a computer instead… It's almost unbelievable in comparison to the previous generation. All that heavy work, people probably got killed at these places and hurt themselves, lost a hand and all kinds of things. I often think about that when I visit these places, about the people who worked there. Could they even imagine that in twenty to thirty years there would be some grown-up kid travelling around taking photos like it was the pyramids of Egypt or something, seeing it as an exciting tourism object?

Sven's attitude to the abandoned limestone quarry echoes the cosmopolitan inter-play between mastery and surrender discussed in Chapter 5 (see also Hannerz, 1992: 253). Urban explorers attempt to take aesthetic and intellectual control of historical places, becoming connoisseurs, while at the same time surrendering to the places' mythological power and ghostly elements. Sven's attitude also resonates with Beck's (2006) cosmopolitan vision of dialogic imagination, referring to how the cosmopolitan self is not only re-imagined through the eyes of the Other, but also potentially undergoes change. When Sven reflects upon his visit to the old Cementa industrial plant he starts questioning his own interest and even sees it as a childish expression of our mediatized times where hard work has been replaced by touristic adventures and interactions conducted through "small cameras in comput-ers". This means that the act of exploring the site as such is interrupted and the gaze is turned back upon the adventurer him-/herself. This moment of *reflexive hesita-tion* (a notion that I will return to below) is closely tied to visual practices, notably photography, whose objectifying and potentially exploitative function may seem at odds with the predominant ambition among urban explorers to pay tribute to the people who once inhabited the places (in contrast to graffiti painters) *and* to dis-tance themselves from dominant forms of touristic practice (such as ruin tourism).

This brings me to the second area of tension, which concerns how urban explor-ers conceive of their *spatial practices in relation to society at large*, and in relation to *hegemonic forces* in particular. Again, the tension found among urban explorers coin-cides with tendencies associated with post-tourism. On the one hand, as already discussed, urban explorers can be seen as *place-political activists*. Their resistance

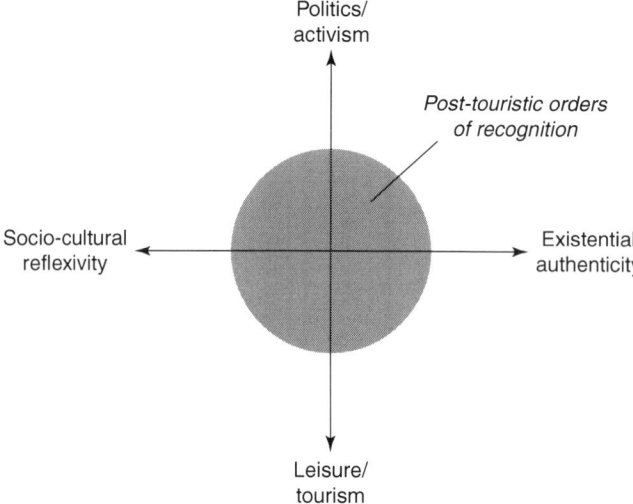

FIGURE 6.1 Post-touristic orders of recognition.

typically concerns commercialized or otherwise enclosed urban spaces, as seen for example in transgressive anti establishment acts of "place-hacking" and "recreational trespassing" (Garrett, 2014a, 2014b; see also Kitchin and Dodge, 2006), which problematize who has the right to the city (see Lefebvre, 1968/1993; Harvey, 2012). As well as this, the counter-hegemonic ethos of urban exploration also attains a broader scope related to its insistence on being different from mass tourism and actively recognizing *Other* time-spaces, as seen in the exploration of abandoned (and not necessarily "urban") industries, hospitals and work places.

On the other hand, the desire among middle-class subjects to be different cannot be uncoupled from their anxiety to distance themselves not just from commercial forces but also from *class fractions below them*. As Munt (1994: 119, italics in original) concludes in his analysis of post-tourism, focusing particularly on the marketing of Third World destinations, the individualistic, emancipatory drive of post-tourism "signals a cultural and social *re*action of the new middle classes to the crassness which they perceive as tourism, and their craving for social and spatial distinction from the 'golden hordes'". Understood as post-tourists, then, urban explorers represent the socio-cultural construction of what Lawler (2005) calls "disgusted subjects". While criticizing what they judge as touristic, however, they operate within an overlapping and *largely leisure-oriented culture of spatial consumption and display*.

These ambiguities, which clearly re-assemble the dominant order of organized self-realization (Honneth, 2004), can also be found in relation to mediatization. Urban explorers commonly express a desire to be unique and resist mediatization (even though other phrases are used to describe this). They want to find places that nobody has visited before (that is, after they were abandoned) or that only very few have access to. They rarely go back to the same place twice. At the same time, as we have seen, the celebration of uniqueness and authenticity is part of their recognition work; it becomes a way for individuals to position themselves as autonomous and respected subjects within the community of urban explorers. While their community-defining practices may stand out as rather odd in relation to the social world at large, they submit to certain codes and principles, what we may call doxa (see Chapter 4), that define what urban exploration is and *is not*. When it comes to communicative practices we may even speak of an "urbex communicational doxa" that pervades the culture of urban exploration. This doxa, in spite of its critical stance towards staged tourism and spectacularized heritage sites, is largely reliant on media technologies. In the following sections I draw the contours of this urbex communicational doxa, focusing on three inter-related stages: *site-discovery*, *camera-work* and *storytelling*.

Site-discovery

It should be stated first up that urban exploration is not an explicitly media-critical movement. Urban explorers are not engaged in media activism or "hacktivist" projects even though they maintain an alternative and largely anti-hegemonic ethos. They use media neither in technologically disruptive ways nor as material for artistic interventions in controlled, spectacularized or otherwise mediatized environments (see, e.g., Elwood and Leszczynski, 2012; Gansing, 2013; Pinder, 2013), at least not in the typical case. Rather, the focus of urban exploration is on *space and its transformations* over time, involving processes that are more indirectly, and in diverse ways, linked to mediatization. Urban exploration thus merges the individual, and middle-class biased, quest for self-realization with an intellectualized engagement with the politics of place (Harvey, 2012). Here media occupy a dual position, functioning as a tool for discovering and exploring places while at the same time being understood as a problematic context and driver of contemporary spatial exploitation and commodification. This leads to the paradoxical situation that while urban explorers go to a great deal of effort to search out places that are either "non-mediated" or located beyond or underneath heavily mediatized environments, they also sustain mediatization processes through their own communicative practices.

This tension is particularly obvious in relation to the discovery of new, unique places to explore. Urban exploration can be seen as a knowledge community. By this I mean that urban explorers continuously seek out new information and develop specialized knowledge about certain places or areas of interest, partly in dialogue with other explorers. Within the Swedish urbex community there are certain key figures who are frequently mentioned in the interviews and described as particularly knowledgeable authorities within their fields. These individuals legitimize certain kinds of knowledge that, along with the achievement of having explored a significant number of sites, constitute a key source of symbolic capital. Accordingly, the search for new sites does not occur at random, even though there are many examples of coincidental discoveries taking place when travelling or just talking to people. Rather, it is framed by intellectualizing discourses that guide the explorer to certain sites rather than others. Newly explored sites are turned into something more than simply new items in the collection of places; they also become pieces of information that contribute to the broader accumulation of historical knowledge. This is not to say that every urban explorer is a hobby scientist; the emotional thrill of spatial adventures is often more important than understanding the broader contexts. My point is that the endeavour of discovering new sites and relating them to other sites is necessarily framed by historical and geographical discourses, which are in turn dependent on mediation.

The Internet in general and dedicated urbex forums in particular play key roles in this process. Not only do online resources provide facts about particular sites and their historical contexts, they also bring urban explorers in touch with one another and enable the exchange of practical information concerning the qualities of particular sites – such as their level of accessibility and whether they are worth visiting – as well as technical questions pertaining to equipment and photographic techniques. The interviews contain many examples showing how these exchanges can lead to further collaboration, such as in joint explorations and giving mutual help and guidance to places of interest. Such interactions, however, are based on trust and are framed by certain codes of conduct in order to ensure that information about valuable sites is not leaked to people with the "wrong" interests (and who are not supposed to be part of the community).

> Especially those who are a bit bigger and who are more serious, they hold onto their information. You might meet a friend of a friend who is interested and you talk about some well-known stuff that you can find in the city and that most people already know about. But those who have been doing this for several years and who are well known and have been active in urban exploration, they don't want to disclose their coordinates because they are afraid that people will go there and destroy. And it's always kind of nice to know more than others know. To be in charge of information.
>
> *Zac – urban explorer in his mid-twenties*

This also means that those explorers who are well-established in the field and have acquired credibility among other explorers are in a better position to gain access to exclusive information. It even happens that they are contacted and given tips about places that could be of interest to them.

Both online and offline information sharing are essential to the formation of the urbex community as a field of status relations. Sharing practices adhere to a regulative pattern, doxa, in which the value of different sites and explorations is defined. For example, the value increases if there has been a considerable amount of work and research involved in locating the site. Visiting a well-known site or just following in the footsteps of other explorers affords little, if any, status. Similarly, contacting somebody one does not know to ask for directions is not considered appropriate. Information sharing should always be based on trust. As shown in the opening extract of this chapter, the highest ambition for those urban explorers who search for abandoned places is to be the first person to enter a site after its abandonment.

The value of such a rare discovery increases the longer the site has been abandoned and the more it has escaped mediation and destruction.

The vision of the urban explorer is closely linked to neo-romantic views of independent mobility through uncharted terrain (cf. Garcia, 2016). Even so, such adventures are framed by the rational utilization of media technologies, as Niklas describes:

> The Internet is good for many things, as I said. Some places we have even found through Google Earth; we've just sat down with it and searched. And books… yes, but in books they are not that concerned with specifying exactly where things are located, if we are talking about abandoned places. There might be a picture added to a map pointing to a crossroads or something, which we have then searched for with Google Earth until we've found a similar looking piece of road. And then we've pasted that map onto the other and that's how we have found new things.

In this description, and in most middle-class settings today, Google Earth and Google Maps belong to the realm of everyday life. Along with other geo-media applications they have become standard tools for looking up directions and gaining pre-understandings of destinations. Among many urban explorers Google Earth is the natural shortcut to the "real adventure" that takes place in the socio-material geography. Such are the contours of a mediatized society, where media technics constitute an increasingly complex, yet invisible interface between self and world (Ihde, 1990; see Chapter 3). As Elwood and Leszczynski (2012: 544) argue, based on a study of activist groups and various grassroots communities, new "spatial media" – including geographical online representations and their mediating technologies – foster knowledge politics that entail "a prioritization of individualized interactive/exploratory ways of knowing […] and approaches to asserting credibility through witnessing, peer verification and transparency".

The logical consequence of such a development is that certain forms of unmediated encounters with spaces, places and people gain a more distinctive value. A concrete expression of this development within the field of urban exploration is that the uses of new media technologies "drive inflation" and simply reduce the number of places that are not yet explored and mediated. While urban exploration was a rather peripheral phenomenon in Sweden until the mid 2000s, the publication of a few significant books (see especially Jörnmark, 2007, 2008) and the parallel expansion of digital camera technologies and social media platforms have turned this leisure activity into something of a well-known phenomenon with a steadily growing number of participants all over the country (albeit still not seen as mainstream). This means that more and more regions have their own groups of dedicated explorers who map out and discuss abandoned places online, spurring the interest further and catalyzing the uses of advanced technologies for spotting new places. Among the interviewees of this study there was little explicit critique of this accentuated mediatization; on the contrary, media are continuously promoted as essential parts of urban exploration. In several of the interviews, however, one can trace concerns that perhaps *too many people* are now entering the field, including those with a more superficial interest in the topic, and that the very landscape of exploration might soon get exhausted, emptied out of places to discover.

Camera-work

A second key aspect of mediatization concerns the appropriation of advanced and user-friendly camera technologies. Several urban explorers have stated that their involvement in the community actually started with an interest in photography. One of the interviewees describes how he once bought a system camera and started taking photos of the things he was interested in, like cars, music festivals and places he visited. His interest in photography gradually merged with an interest in discovering abandoned places, mainly because these were the kind of photos he found most exciting and rewarding to take. For other interviewees, photography became an indispensable part of urban exploration because of the need to document sites in order to remember them and show them to others. One interviewee with a particular interest in hidden and/or abandoned military sites mentioned that he initially explored these places without any camera equipment, but found it boring to always be waiting for his friend who regularly stopped and took photos. After a while he also bought a camera and got hooked on photography.

While the primacy of photography varies between different explorers, as do their levels of professionalism, a common feature is that photo quality is considered highly important to the overall practice. Furthermore, taking good photos has become more important due to the growing accessibility of digital cameras and the extensive online circulation of images. While the continuous public exposure of new pictures means that people feel a pressure to improve, it has also become cheaper and easier to develop one's skills and to gain inspiration from others.

> Before the Internet, and before the digital camera, I usually took some transparencies, one role perhaps, not more. It was a matter of documentation. But now… It costs me 100 crowns [around 10 euros] a year to keep this website so it's no big thing. That has also become a kind of documentation.
>
> *Sven*

> We've been to transformer stations where trains were stored in the past, we've been to many such places. Now the exciting thing is to find one that is different and where there are new things to shoot and so on. Preferably in better condition. Or to be able to find better angles and spots to take photos from. To learn more all the time. If all rooms look exactly the same it's not as much fun.
>
> *Roger – urban explorer in his mid-twenties*

These examples show how photographic practices, and more broadly the dialectic of mediatization, evolve through the interplay between orders of technology and orders of recognition. While Sven's story highlights the loosened exclusivity of taking photos, it also shows how digitalization has altered the communicational doxa of urban exploration, which now prescribes extended forms of documentation and public exhibition as a more or less mandatory activity. It even happens that urban explorers visit particular places, sometimes located far away from their home regions, in order to try out new media equipment under the right conditions. This development can be seen as an accentuation of mediatization, even though photography has been an integral element of urban exploration all along. At the same time, as shown in the second extract, the doxa of urban exploration feeds into photographic practices and spurs the desire to develop one's skills to a higher level. When one's efforts are exposed to a broader audience of like-minded peers, the underlying ambition to document one's explorations is transformed into a more complex and reflexive practice. It becomes important to show *originality* as well, in terms of both *which* sites one depicts and *how*. As Roger puts it, "it might be two years since you took the pictures [on the website] and they may look like a disaster, but then you can go back and take new and more varied pictures".

To some extent, personal websites of urban explorers function as displays of photographic progress, epitomizing the merger of camera-work and recognition-work. While technology has made it easier to produce and share an excess of photos, there are doxic rules that stress quality rather than quantity. Taking quality photos is not just a way of gaining status within the field of urban exploration. It is also associated with the normalized endeavours of mobile middle-class subjects to acquire distinctive markers of taste. The notion of "quality" and "taste" here points to a process of cultivation whereby the meaning of the photography alters from being a means of documentation to becoming an end in itself; that is, *a form of art*. This transformation, which is reflected in Roger's description of his older pictures as aesthetic "disasters" (see above), corresponds to the classic division between "vulgar" and "noble" tastes and symbolizes an ascending social trajectory within the hierarchy of class relations (see Bourdieu, 1965/1990). Similarly, when other urban explorers describe their photographic practices they often stress the importance of finding new angles, using only natural light and/or authentic light sources (instead of automatic flashes), and so forth. The communicational doxa is thus marked by aesthetic standards that resemble the taste patterns of the dominant classes and lead to the establishment of explicit structures of classification.

The significance of aesthetic recognition, by means of photography, also shapes the onsite textures of urban exploration. While there are good reasons for urban explorers not to visit abandoned or concealed places on their own, because it might be both difficult and dangerous, photography reinforces the individualistic ethos of spatial adventuring. Accordingly, the number of participants is usually kept to a minimum, enabling mutual support while not disturbing the creative processes or the possibilities for gaining authentic experiences.

> Often you have to help each other. Like if someone is about to take a photo in a big room underground and it's completely dark. Let's say that you're

about to take a photo and you need a lot light, then you have to try different angles and my friend might ask me to stand on one side and light up a certain angle. So you have to help each other.

Zac

But there can't be too many of you either, because then there will be people in the pictures all the time. It's impossible to take photos if there are, like, ten people around. So three is like the perfect number when it comes to photographing abandoned places. […] But those that I'm photographing with, they are also good photographers, and we're usually doing it in a very organized manner.

Niklas

Many urban explorers develop strong partnerships, meaning that they organize their journeys and visits in couples or groups of three. These small constellations can be seen as communities of trust and knowledge. As described in the last quote, exploring a site in a larger group would make it difficult to get good photos, not just because of general disturbances but also because of the difficulties of grasping the spatial atmosphere and "sensing the ruin" in an immersive way; that is, through open-ended encounters with all sorts of materialities (Edensor, 2007: 217). One of the explorers explained that when he *once* participated in a group of eight explorers "the whole place turned into a kindergarten". A few of the participants had not behaved in the respectful and courteous manner that was expected and even disturbed the photographic process. However, the comparison with a kindergarten speaks of a more fundamental cultural concern than just practical problems related to photography. It also resonates with the general disgust with mainstream tourism, including pre-organized trips, crowds and inferior photographic tastes. The artistic ambitions of urbex photography are thus part and parcel of the post-touristic ethos of the middle classes . While photography shapes the way in which space is appropriated and sensed, it is in itself shaped by the cultural desire for distinction.

This distinction is not necessarily achieved with any ease, however. Besides the fact that photography in itself, in order to become something more that just documentation, demands improved skills and a certain feel for originality that cannot be easily appropriated without a successive negotiation of habitus, the complexity of the space itself often presents a barrier. While several of the interviewees vividly describe the thrill of taking "cool pictures" that depict unique scenes – notably the recurring motif of places that seem to have been left in haste – there are also stories of emotional ambiguity tied to the conflict between photographic objectification and what we may call "historical compassion" associated with the social narratives of a place. This applies particularly to abandoned sites marked by human activity, life and death over a long period, where photography accentuates the tension between spatial mastery and surrender. As we saw in Sven's story, where the history, solidity and volume of the abandoned limestone quarry made him feel like a little schoolboy, or, alternatively, *a tourist*, with a camera, photography can trigger a moment of self-reflection. Taking up the camera and starting to shoot becomes a sign and a reminder of one's own privileged situation and how one's social trajectory is linked, directly or indirectly, to the lives of people of the past.

Photography thus contributes to a peculiar kind of hesitation, even guilt, in relation to the sites. In the interview material there are few obvious examples of the possessive or exploitative tendencies reported in some previous studies (see, e.g., Bennett, 2011; Mott and Roberts, 2014). While exploration as such can be interpreted as the masculine desire to discover and map out new territories, there is also a recurring emphasis in the interviews we conducted on leaving the sites exactly as they were found. This attitude corresponds to the post-touristic doxa of existential authenticity, which in turn demands that photos represent things *as they currently are* rather than in a staged manner replicating the past or reproducing commercialized formats of "ruin porn" (Millington, 2013) or picturesque "ruinscapes" (Tegtmeyer, 2016). When materially untouched, the meaning of the site remains open to interpretation and creative reconstruction by means of other media.

Storytelling

How is it possible to maintain a cultural community that is at the same time critical of mediatization and inherently dependent on media for the documentation and sharing of spatial experiences? As we move to the last realm of the urbex communicational doxa – the construction, sharing and spreading of urban images and stories – these inner tensions are accentuated. What appears are the contours of a complex structure of feeling, marked by dissonant experiences, even translatable into a social sub-field containing multiple competing positions.

In her analyses of Danish urban explorers Klausen (2012, forthcoming) describes urban exploration as a subversive culture of mediatization (see Hepp, 2013), defined by the ways in which it "transgresses everyday codes of where to go and what to do in the city" (Klausen, 2012: 561). At the same time she criticizes the romanticized view of urban explorers – typically advocated by researchers who are themselves associated with the urban exploration movement (e.g., Garrett, 2014b) – as oppositional groups involved in a counter-hegemonic struggle. Having analysed a handful of cases from ethnographic "go-alongs" (i.e. interviews conducted onsite), including with the quite well-known duo CphCph, Klausen concludes that the spatial practices of urban explorers are inherently mediatized in the sense that their spatial expectations and practices are guided by well-established and widely circulated standards of what typical "urbex sites" should look like and how explorers are supposed to present their identity. The genre known as the "hero-shot", depicting an independent (typically male) explorer on a rooftop high above the city or in any other dangerous and more or less inaccessible place, contributes especially to the spread of a particular "urban exploration imaginary" (Klausen, forthcoming; see also Mott and Roberts, 2014). Such images flourish on publicly accessible social media sites like Facebook, YouTube and Flickr, from which they gradually spread to a wider audience and eventually even influence dominant cultures of advertising (e.g., the 2014 Nike campaign for *All Conditions Gear*). This means that the transgressive attitude of urban exploration is translated into an aesthetic pattern that ultimately contributes to the commodification of both the city and the urban explorer identity.

Klausen's analysis highlights how the spatial practices of urban explorers are sometimes adapted to dominant "social media logics" (see Van Dijck and Poell, 2013), sustaining the drive towards increased popularity or what Jenkins, Ford and Green (2013: 4) call spreadability: "Spreadability refers to the technical resources that make it easier to circulate some kinds of content than others, the economic structures that support or restrict circulation, *the attributes of media text that might appeal to a community's motivation for sharing material*, and the social networks that link people through the exchange of meaningful bytes" (italics added). In the case of CphCph the explorers even turn out as cultural entrepreneurs who are in the business of generating maximum popularity and sales for their photos, books and videos. The agents speak explicitly about their media practices in terms of spreadability and mention the indispensability of Facebook for reaching out efficiently to a broad audience. While the spread of this urban exploration imagery might contribute to a broadened sense of public participation and "ownership" of the city – in the sense that "ordinary people" can get the chance to see otherwise hidden parts of it and gaze at it from new perspectives, sometimes even sharing or commenting on historical anecdotes or facts from certain places – it is also part of the hegemonic construction of the city as mediatized consumption machinery, a creative node within the late modern economy of signs and space (Lash and Urry, 1994).

Klausen's analysis thus presents an important counter-argument to the view of urban explorers as a counter-hegemonic sub-culture. It should be noted, however, that publically acclaimed actors like CphCph cannot be seen as typical examples of the broader urban exploration community. As Klausen also notes, there are strong internal forces against the mainstreaming of urban exploration, grounded in the fact that many explorers want to go on with their activities in secret (relatively speaking) and thus have an interest in *not making the hobby too popular*. The Danish urbex community, for example, moved their online activities from Google Maps, where they used to share information about different locations, to a private forum on the Internet in order to protect abandoned places from unwanted visitors and to *distinguish themselves from other sub-cultures* (Klausen, 2012: 571). The commodification processes and image inflation that follow from the kind of connective media strategies enacted by CphCph tend to make the whole culture of urban exploration less distinctive. They also undermine the activist ethos and the ideal of existential authenticity discussed above.

The Swedish material provides a more diverse view of the urban exploration community. While several of the interviewees use various online resources to circulate their experiences and get in touch with each other, the ethos of independence and restricted circulation predominates. Besides the fact that commercial mainstream channels like Facebook are often met with a certain degree of scepticism, the interviews actualize a division between spreadability and *storytelling*. The desire to reconstruct and share a trustworthy story about a place is in many cases as important as the desire to discover it, and clearly more significant than exhibiting oneself.

> I don't think this has anything to do with exhibitionism. There is today some kind of Facebook sickness and of course people only tell the good things, like, "Look, what I nice breakfast I had", or, "Look, now we're at the Grand Hotel having breakfast"… It's a very sanitized world… I'm also selective of course, I tell the story I want to tell, more or less consciously, with the difference that I have my own webpage while others have Facebook or Instagram. […] I start out with facts and then I have to fill in some gaps with my own fantasy. I say on the webpage that there is no guarantee that these are exactly the facts.
>
> *Sven*

Sven's statement ties in with the understanding of dominant social media as intertwined with the ethos of mainstream tourism in which only certain sights and experiences qualify as sharable or spreadable (see Munar and Jacobsen, 2014; Albrechtslund and Albrechtslund, 2014). From such a "mediatization sceptical" viewpoint social media symbolize a standardized and socio-culturally sanitized leisure culture marked by lack of independence and creativity. Keeping an individual webpage becomes a marker of distinction, an indication of keeping-at-a-distance from commercialized platforms and a more intellectualized approach to the representation of urban exploration. Among explorers like Sven, the accent is more on the vertical (or time-biased) exploration of historical layers of social life than on the horizontal (or space-biased) adventuring into dangerous areas and search for spectacular vistas.

One can speak therefore of *different fractions* of the urban explorer community, or positions within the field. Different fractions are interested in different types of sites and practise urban exploration in partly different ways. While some are more akin to leisure-oriented and self-expressive movements like parkour, others are marked by a civic (or activist) ethos and engagement in alternative productions of place and heritage (see also Figure 6.1). In the latter case, status is not gained through mediated popularity but rather through the acquisition of intellectual capital within alternative knowledge systems. The sense of encountering a place whose story is not already fixed or packaged, but rather has to be re-assembled and narrated for the first time, is part of the thrill. While it might be difficult to be the first to discover a particular site, one can still be the first to *tell its story*, or to tell it in a new way.

These positionalities highlight certain variations within the urbex communicational doxa. They also testify to the dialectical nature of mediatization as a cultural force. The ways in which different actors relate to mediatization, and to the culture of connectivity (Van Dijck, 2013), become an integral part of their recognition work. Most often, however, urban explorers belong to neither of the extremes, but must continually negotiate different elements of doxa. As suggested above, this can be described as a moment of *reflexive hesitation* tied to the questions of *whether* and *how* to represent a certain site; that is, *whether and how to mediate*.

Conclusion: Mediatization as a force of reflexivity

Urban exploration is far from a mainstream activity. This chapter has argued that urban explorers can be categorized as a certain type of *post-tourists*, eager to distinguish themselves from tourism and other expressions of commercialized and mediatized leisure culture. The general purpose, or ethos, of urban exploration can be summarized as five key elements: (1) the felt need to document abandoned or hidden places; (2) the thrill of accessing forbidden or marginal places; (3) the desire to discover and experience authentic, non-exploited places; (4) the drive to create original stories and alternative aesthetic (especially photographic) representations of place; (5) the place-political ambition to problematize the control over places and cultural heritage. More or less the same elements have been identified in previous research (e.g., Dodge and Kitchin, 2006; Klausen, 2012; Robinson, 2015). At the same time, it is obvious that urban exploration is a more heterogeneous movement than one might first think. The five elements are accentuated to varying degrees by different explorers, leading to different patterns of spatial and communicative practice – balanced between socio-cultural reflexivity and existential authenticity – and different positionings in relation to the hegemonic structures of society (see Figure 6.1).

If we think of urban exploration as a cultural sub-field we can then identify competing positions, defined along Bourdieusian lines of division, between, on the one hand, popularity and visibility (corresponding to economic capital) and, on the other hand, intellectual and artistic status (corresponding to cultural capital) (see, e.g., Bourdieu, 1979/1984). Communicational doxa fluctuates according to these dimensions and orders of recognition. While the intensity of competition between various fractions of the urban exploration community can be described as low – based on the available material, and especially compared to the explicit rejection of other forms of urban leisure (e.g., graffiti and ruin-tourism) – the analyses show that spatial and communicative practices themselves are burdened with a particular kind of ambiguity, what I call *reflexive hesitation*. For example, many explorers speak of the difficulties in deciding whether to publish images accompanied by information that might reveal the geographical position of a particular site. This is partly a matter of protecting the site from criminal damage or otherwise destructive elements that might make it less appealing to other urban explorers. But it is also a matter of recognizing and re-affirming the lives of the historical Other, of not breaking the spell of a certain site.

We can recall Edensor's (2005: 847) arguments regarding the alternative forms of knowing that the materialities and imaginaries of ruins evoke, forms of knowing that are based on "empathetic and sensual apprehension" rather than intellectual knowledge. "This kind of remembering implies an ethics about confronting and understanding otherness (here, the alterity of the past) which is tactile, imaginative, and involuntary. It cannot be imperialistic because it must be aware of its own con-tingent sense-making capacities, and, allowing external interruption and sensory invasion, is porous and refuses fixity". This is the opposite of spatial exploitation and control. However, the primacy of representational practices among urban explorers, combined with the socially normalized expectations to *connect* and *spread* media material to broader audiences – that is, the culture of connectivity (Van Dijck, 2013) – challenges this type of impressionistic and non-mediated appreciation of the past. The reliance on media causes emotional and ethical ambiguities that urban explorers must handle reflexively throughout the different stages of exploration and must ultimately interweave into their own recognition work.

These observations confirm how mediatization works as a *force of reflexivity*. They also suggest that urban exploration represents a rather distinct, albeit not entirely homogenous, structure of feeling. The community rests upon a shared socio-spatial ethos that emanates from certain ways of experiencing contemporary society, and, on the practical level, translates into *continuous moments of reflexive hesitation, reinforced by mediatization.*

How can we further position and explicate the significance of such a structure of feeling within the broader context of mobile lives (the hegemonic structure of feel-ing) and mediatized culture and society? It should be noted, first of all, that urban exploration is marked by a middle-class bias, which also implies that the movement has links with a range of other socio-cultural phenomena that share similar forms of sensitivity to cultural change. As I discussed at the beginning of the chapter, such phenomena belong to the broader realm of post-tourism and include various types of adventurous activities and intellectualizing ways of consuming and representing space (see Urry, 1990; Munt, 1994). What they share is a reflexive ethos that prob-lematizes dominant structures of cultural classification through the exploration of new expressions and experiences, ultimately contributing to the realization of these groups' upwardly mobile life trajectories. While it is difficult to find any *explicit* examples of such social strategies among the urban explorers interviewed for this study, which is also part of the social logic, their ways of discussing the relationships between their own interests – especially their taste for certain kinds of places and modes of representation and communication – and the leisure activities of Others are marked by distinction.

Still, it is difficult to say, like Munt (1994), that post-touristic structures of feeling reside *within*, and directly reinforce, hegemonic orders of spatial organi-zation and escalating mediatization. Rather, as in the case of gentrification, the socio-cultural "in-betweenness" of post-tourism implies that individual actors are torn between hegemonic and counter-hegemonic orders of recognition (see Chapter 7). Activist elements compete with desires to aestheticize and regenerate. Accordingly, even though urban explorers strive to leave sites as they are, even keeping them secret, there are many examples from cities around the world of urban exploration sparking off processes of gentrification through the signposting of new areas for middle-class elective belonging. This is captured well in a 2014 British newspaper article:

As more and more international capital flows into London, so public space is increasingly eroded — it's just too valuable for us ordinary folk to paddle about in any more. The place-hackers demonstrate what happens when property rights of any sort are challenged: the law comes down on you like a ton of London stock bricks. But one of the many ironies in this imbroglio is that almost all of the sites they entered have now become valuable real estate: it's as if the place-hackers' derring-do made these abandoned bunkers and sealed-up tunnels appear sexier and more marketable than heretofore.

Self, 2014

Urban exploration can thus be singled out as a force of socio-spatial pre-mediation (see Grusin, 2010; Jansson, 2013a), marked by a certain type of spatial sensitivity that anticipates broader movements of middle-class mobility and entrepreneurship. Since the 1990s such transformations have occurred on a large scale in many post-industrialized cities, most clearly epitomized in the de-populated, decaying and partly redeveloped city of Detroit that has turned into a veritable adventure space for urban explorers, ruin tourists, artists and other types of alternative groups whose amalgamated spatial and communicative practices gradually turn certain derelict areas into epicentres of creative energy. As Slager (2013: 36–39) found in her study of ruin tourism in Detroit, the circuits of representation through which such transformations occur are difficult to govern as soon as a given imaginary structure comes to dominate and becomes a source of spatial phantasmagoria. In a culture of connectivity the representational practices of (more or less) counter-hegemonic groups, such as urban explorers, become ever more intertwined with the normalizing practices of travel journalists, commercial photographers, designers and other cultural intermediaries, whose ethos basically gravitates around the same structure of feeling, *mobile lives*. In the next chapter, as we turn to the interplay between mediatization and gentrification, we will look more closely at these questions of symbolic and spatial power.

7

MEDIATIZATION AND GENTRIFICATION

There are no precise rules or common themes that can describe what constitutes "middle-class taste", "middle-class mobility" or "middle-class media". In Chapters 5 and 6 I depicted lifestyles that on the surface do not seem to have much in common. Elite cosmopolitans are white-collar professionals who travel between destinations prescribed by their companies or organizations. They do not, and cannot, engage too deeply with their current places of residence, since they know that they will probably stay for only a limited amount of time. Urban exploration, in turn, is a post-touristic leisure activity marked by open-ended mobility that involves searching out and visiting places because of their raw authenticity and aura of decay and/or remoteness. Places are consumed in an intense manner, but left untouched. There are also significant, and media-enhanced, variations *within* these groups. Expatriate lifestyles differ not only according to location and social field but also depend on, for example, gender and family structures and career stage. As Polson (2016) reports from her fieldwork among expats in three different settings (Paris, Singapore and Bangalore), the appropriation of geo-social media tends to reinforce the stratified nature of this emerging "global middle class". Similarly, urban exploration (and post-tourism at large) can be analysed as a socio-cultural sub-field in which spatial and communicative practices are differently attuned depending on the positions of various agents.

So what is the point in comparing such diverse expressions of mobile lives, besides the obvious fact that they are mobile in different ways? As I stated in the introduction to this book, and elaborated in Chapter 4, middle-class mobilities are interesting precisely because of their multi-faceted and relatively dynamic nature, which in turn can be related to their underlying habitus. Rather than being shaped by strong traditions or socioeconomic restraints, middle-class mobilities typically constitute an *aspirational space*, one that is both contested and hegemonic. Broadly speaking, while the middle classes (and especially certain class fractions) actively challenge and negotiate cultural norms and dominant taste patterns in order to improve and legitimize their own status, as reflected for instance in the desire for novelty, authenticity and difference, they set new social standards and horizons of expectation, including the normalization of certain forms of privileged mobility and certain ways of appropriating space and place, that affect broader strata of society. As such, expatriate lifestyles and urban exploration are phenomena that articulate self-realization as a core value of middle-class life. When these aspirational values materialize through everyday practices they also bring out the ambiguous role of mediatization, one that oscillates between enhancing the capacity to manage mobile lives and escalating levels of technological and social dependence.

What also unites elite cosmopolitanism and post-tourism, and sets the scene for the analyses of this chapter, is that they involve practices that affect the *production and coding of space* in class-specific ways. According to Lefebvre (1974/1991: 47–48), "a spatial code is not simply a means of reading or interpreting space: rather it is a means of living in that space, of understanding it, and of producing it. As such it brings together verbal signs (words and sentences, along with the meaning invested in them by a signifying process) and non-verbal signs (music, sounds, evocations, architectural constructions)". As we learned from the analysis of urban exploration in the previous chapter, there is a tendency for the circulation and aestheticization of alternative sites of spatial consumption to go hand in hand with the commercial exploitation and regeneration of these same locations; for example, former industries, shipyards and warehouses. Urban exploration can thus be seen as both a driver of, and a reaction against, such recoding processes whereby novel spaces of middle-class attraction are produced. Similarly, my fieldwork in cosmopolitan Geneva exposed a relatively segregated city in which a distinct line of division was drawn between the Right Bank (in relation to the river Rhône), mainly inhabited by the mobile "international community", and the Left Bank with its more locally rooted Swiss communities. However, the Right Bank not only holds the fancy hotels and institutions associated with the United Nations and other international organizations (as well as banks and corporate headquarters), but also the socio-culturally mixed neighbourhood of Paquis with its red-light district and a great variety of restaurants, clubs and markets. Some of the expatriates I interviewed had decided to live in Paquis not just because of the economic advantages (Geneva is a very expensive city), but also because they wanted to get out of the "expat bubble" and encounter a more diverse group of people in their day-to-day lives.

This brings us to the question of *gentrification*, which literally means that the "gentry", that is, people of good social standing, come to dominate and reshape spaces that were previously populated by, or in other ways classified by, less privileged groups, notably the working classes (Glass, 1964). Gentrification raises a critical research agenda that ultimately deals with matters of *displacement*; that is, how people with less capital are forced to move from their neighbourhoods as these become socially upgraded and more expensive to live in. The forms of middle-class mobility explored in this book provide good insights into the socio-cultural mechanisms of such changes. In addition, the current analysis will highlight the relationship between gentrification and mediatization – an issue that has so far been given little attention in research.

This chapter thus builds upon and extends the analyses already presented. It is obvious that urban exploration and expatriate lifestyles are in many ways related to gentrification, even though they do not point primarily to long-term settlement and home-making, and that media play important roles both culturally and technologically. In addition, this chapter includes analyses based on interviews with relatively privileged individuals who have settled (and are thus newcomers) in two different areas in Sweden: first, the recently regenerated Western Harbour area in the city of Malmö (similar to Docklands in London and other coastal cities), and, second, the provincial municipality of Arvika. As stated in Chapter 1, these interviews are derived from two different research projects covering a ten-year period between 2003 and 2013 and deal with urban and rural/provincial gentrification respectively. They also deal with two very different media landscapes (on account of the time-span). What they have in common, however, is that they provide a critical, bottom-up perspective, an inside view, of how different media resources sustain creative place-making processes, especially through the circulation of alluring images of urban diversity and countryside idylls, while at the same time complicating people's relationships to these very same places.

The chapter contains four sections. In the first section I provide a brief overview of the gentrification debate and the lack of research within it concerning the role of media and communication. In the second section I theoretically assess the relationship between mediatization, gentrification and *elective belonging*. The latter concept was introduced by Savage et al. (2005), inspired by Bourdieu's sociological work, and refers to the socially structured logic according to which relatively well-off people find and create home-places that "fit their tastes". I argue that elective belonging in the current context can be understood as the particular structure of feeling, an articulation of mobile lives, that characterizes the settlement processes of mobile middle-class groups. It describes the emotional textures that underpin both urban and provincial gentrification processes (or at least one side of these processes, since other groups may experience marginalization and displacement) and is here applied as the analytical tool through which we can gain an experiential view of the relationship between mediatization and gentrification.

In the subsequent two sections I study two inter-related tensions of middle-class home-making projects in order to once again unpack the dialectic of mediatization. I start with the middle-class orientation towards *cultural difference* and self-transformation that establishes settlement processes as adventurous, and mediatized, biographical projects. At the same time, however, gentrification processes, reinforced by media, have a tendency to foster segregation and exclusion. I then discuss the creative form of *topophilia*, or "love of place" (Tuan, 1974), that distinguishes middle-class home-making projects, and the ambiguous ways in which media technologies form part of these emotional investments in place. I point especially to the indispensability of media among middle-class counter-urbanizers for maintaining a sense of *exitability* in provincial areas. This also brings me, by way of conclusion, to the question of social power and the significance of mediatization as a *mobility regime*.

Putting media into the gentrification debate

The appropriation of new media has been essential for the construction of home and belonging throughout the modern era, especially within middle-class settings. Several decades ago Williams (1974) discussed the characteristics of media-saturated, suburban middle-class lifestyles in the post-war era (see Chapter 2). Similarly, Spigel (1992) has pointed to the centrality of television in realizing the American ideal of middle-class, nuclear-family life. While the emergence of new media forms has led to the reconfiguration of these cultural patterns, the realization of contemporary middle-class trajectories is still dependent on media. Media continue to generate dreams and visions of the home as a place of security, comfort and self-realization. They are also part of the socio-material textures of the home-place, symbolizing achievement and status. Media anchor and connect individuals and groups within and across different spaces, making the boundaries of the home-place increasingly negotiable (see Christensen and Jansson, 2015: Chapter 5).

We might then also assume that the *spatial consequences* of middle-class mobility have something to do with the expanded affordances of media. Why is it, for example, that certain transitory places – whether we speak of former industrial areas, run-down inner city neighbourhoods or rural backwaters – become attractive to different fractions of the middle classes, who settle there and make these places less available to less privileged groups? While this phenomenon cannot be ascribed to mediatization alone, it is clear that media are in many ways *culturally and materially embedded* in gentrification processes. In spite of this, an overview of the literature on gentrification reveals that the role of media in these processes is both under-researched and under-estimated. Apart from a few observations concerning the general celebration of gentrification in dominant media outlets (e.g., Slater, 2006) and a number of analyses of news media coverage of ongoing gentrification processes (e.g., Jansson, 2005; Zukin et al., 2009; Gin and Taylor, 2010), there are no detailed studies or research projects addressing the broader cultural-material significance of an increasingly networked society in relation to gentrification. Even in her much-cited overview of gentrification research, Lees (2000) does not mention the media – not even in her list of research gaps. And in a more recent summary, Doucet (2014) discusses the state of gentrification research without identifying, or calling for, any studies on media and communication.

This lack of research is surprising, especially since the dominant perception of gentrification "is no longer about rent increases, landlord harassment and working-class displacement, but rather street-level spectacles, trendy bars and cafés, i-Pods, social diversity and funky clothing outlets", as Slater (2006: 738) puts it in a critical account of the field. Since the coining of the term in the mid-1960s (Glass, 1964), gentrification has come to signify an increasingly complex set of economic, social and cultural transformations that extend well beyond the mere residential displacement of working-class populations due to the socioeconomic upgrading of neighbourhoods. In addition to the critical and structurally oriented view of an ongoing, ideologically orchestrated middle-class colonization of attractive urban spaces, ultimately understood as a "global urban strategy" aligned with the neoliberal project (see, e.g., Smith, 2002; Slater, 2006), a number of studies have broadened the scope and contributed to a more nuanced view of the socio-cultural mechanisms behind gentrification. These include Zukin's (1982) ground-breaking study of the role of artistic creativity and capital in the recoding of run-down post-industrial districts; Ley's (1996, 2003) work on the new middle classes and so-called cultural intermediaries as mediators between cultural and economic capital in urban change; Phillips's (2002, 2004, 2005, 2010) work on rural gentrification as a multi-faceted articulation of counter-urbanization; Wyly and Hammel's (1999) analysis of the role of "financifiers" in the gentrification of attractive neighbourhoods in American post-recession cities; and a number of analyses addressing the role of gentrifiers in the globalization process (see, e.g., Rofe, 2003; Butler and Lees, 2006), just to mention a few examples. But the media remain a blind spot.

Nonetheless, these studies (along with many others) provide an important framework for considering the significance of mediatization. Above all, they highlight the fact that gentrification should be understood as a gradual shift that looks different in different time-spaces and involves different actors and stakeholders during different stages. There are clear differences, for instance, between gentrification processes led by policy makers and financial investors, as seen in large regeneration projects tied to former industrial land (e.g., Davidson and Lees, 2005; Jansson, 2005) – a type of transformation that is today most often considered as part of gentrification even though it does not entail the direct displacement of residents (see Doucet, 2014: 126) – and the alternative and open-ended forms of spatial change linked to the more organically evolving in-migration of artists and cultural intermediaries attracted by low rents, large spaces and social mixes (Caulfield, 1994). As Lees (2000) points out in considering how gentrification processes (as well as gentrification research) evolved during the 1980s and 1990s, the multi-faceted nature of gentrification also creates drastically different relationships between gentrifiers and place.

> The whole concept of urban community is in transition, the financifiers' ties to the community, to the neighbourhood, are much weaker than those of the gentrifiers of old. The financifier has a much less deeply rooted relationship with his or her neighbourhood – as with the highly mobile capital they work with, these super-gentrifiers are more mobile too – their identity is arguably more fluid than rooted. As such, the term gentrifier may not even be appropriate for these new, well heeled renovators, these *super-gentrifiers* who have displaced sweat equity by employing their own architects, interior designers and builders.
>
> *ibid: 402, italics in original*

Lees points to an interesting tension between mobility and rootedness that also has ethical and political implications. Much of the critique of gentrification has revolved around the harsh economic forces behind spatial transformations, including retail and office spaces as well as housing, that have not only led to the displacement of inhabitants but also ignored and demolished the authenticity and social history of these places. While the mobility of early gentrifiers, those who first discovered the potential of poorer neighbourhoods or otherwise stagnant spaces, tended to be accompanied by a deep-seated interest and engagement in the particular locality, a willingness to invest in place through "sweat equity", those who arrive later in the process or are simply part of a policy-driven make-over project will only see the area's outer shell and, at best, the aestheticized remnants of bygone local communities. The latter form of mobility *may* involve a strong sense of spatial attachment and "pioneership" on behalf of the settlers, but it is bound to be socially segregating and detached from local history. Gentrification processes thus tend to follow a common trajectory, from social mixing, emancipatory visions and alternative lifestyles (e.g., Caulfield, 1994; Ley, 1996) to homogenization, exclusion and commercialism (e.g., Butler, 2003; Butler and Lees, 2006; Zukin, 2008).

As a consequence, gentrification has caused much social turmoil and various forms of political protest ever since the post-war era (see, e.g., Osman, 2011). Such conflicts can be seen as extensions and accentuations of the tension between place-political activism and urban entrepreneurialism that we also discerned in the field of urban exploration (see Chapter 6), and that ultimately align with more general battles within capitalist society. If we are to understand the role of media, and, by the same token, if we are to understand the ways in which gentrification affects mediatization, we have to differentiate between contexts and take into account the multiple temporalities that structure these processes. For example, it is not a far-fetched assumption that the kinds of media practices that accompany the lifestyles of entrepreneurial super-gentrifiers (Lees, 2000, 2003) will differ significantly from those found among more aesthetically oriented artists and cultural intermediaries in run-down areas. David Ley (1996, 2003) has spelled out these dynamics with great clarity in relation to his ethnographic fieldwork in Canadian cities.

> "An old area, socially diverse, including poverty groups" can be valorised as authentic, symbolically rich and free from the commodification that depreciates the meaning of place. For the aesthetic disposition, commodified locations, like commercialised art, are regarded as sterile, stripped of meaning: "there's nothing for me there". The suburbs and the shopping mall, emblems of a mass market and a failure of personal taste, are rejected. The related but opposing tendencies of cultural and economic imaginaries reappear; spaces colonised by commerce or the state are spaces refused by the artist. But, as scholars know, this antipathy is not mutual; the surfeit of meaning in places frequented by artists becomes a valuable resource for the entrepreneur.
>
> *Ley, 2003: 2535, quotations from interviews with*
> *early gentrifiers in the Vancouver area*

Again, as in my study of post-tourism (see Chapter 6), questions of autonomy and authenticity (or lack thereof) are brought to the fore. The gentrification of urban space, ranging from the management of built environments, through policies affecting social life, to the textures of art, media and other forms of communication, seems to encapsulate precisely those cultural ambiguities that haunt middle-class mobile lives. Other researchers (e.g., Phillips, 2005; Hines, 2012; Abrams et al., 2013) have observed similar tensions in the context of rural gentrification, or counter-urbanization, where some in-migrants appreciate above all the authenticity of their properties and make great efforts to preserve as much of the uniqueness as possible (including cultivated land and built environments), while others are more profit-oriented and willing to transform the countryside in ways that attract tourists and other stakeholders of the post-industrial "experience economy". As Ley (2003) argues, gentrification can thus be approached as *a social field* that integrates the tension between different forms of capital as well as between hegemonic and counter-hegemonic forces.

Accordingly, gentrification constitutes another field in which we have a good chance of further coming to terms with the dialectic of mediatization. But how are we to do this? Gentrification as a concept refers to the social re-structuring of space and cannot be used as a tool for analysing the experiences and negotiations of mediatization in everyday life. Rather, we need to identify an underlying ethos, or structure of feeling, that directs these mobile lives and through which we may study the combined experiences of gentrification *and* mediatization in different settings. A particularly useful perspective here, I argue, is *elective belonging.*

Elective belonging as a mediatized structure of feeling

In their book *Globalization and Belonging* Savage et al. (2005) discuss how residential mobility patterns are bound up with classified structures of taste and thus lead to the territorial clustering of more or less like-minded social groups. This spatialization of class relations, they argue, may also lead to tensions between those who arrive as new settlers to a place and those who are already in that place and suddenly have to witness the transformation of familiar environments. Savage et al. thus identify two competing narratives of belonging in this scenario: "nostalgia" and "elective belonging". While the former narrative describes a sense of lost social community, the latter corresponds to the trajectories of the relatively privileged middle classes, who settle and become highly vested in a place in which they may have very few, if any, previous connections. In narratives of elective belonging, as Savage (2010: 118) puts it in a subsequent article, landscapes are appropriated by the newcomers and "construed as a destination on a personal map, a landmark on their own personal journey". At the same time, elective belonging cannot evolve in relation to just any place, but has to embody the classified tastes of the in-migrants (see also Benson, 2014). Such places as "faceless suburbs" and "generic villages" are dismissed by the aspirational middle classes, who look instead for "particular places" with their own identity and "aura" (Savage, 2010: 117).

Gentrification is made up of a multiplicity of stories of elective belonging. This is not to say that gentrification is merely the sum of many individual narratives or the structural outcome of middle-class agency. As discussed above, there are also strong economic, political and ideological forces at play in these spatial transformations. However, understood as an aesthetic and ethical disposition striving to claim beauty and a meaningful relationship to place, I find elective belonging a good entry-point for grasping the relationship between gentrification and mediatization from within. My assertion is that elective belonging constitutes a distinct structure of feeling (a subordinated articulation of *mobile lives*, which I see as the hegemonic form) that characterizes middle-class residential mobility. Elective belonging means that individuals or groups have found *the place* of their taste and made the decision to realize, and adapt, their identities and lifestyles through the material and symbolic production of a new domestic locale. Obviously, this is much more than an economic and social investment. It is a deeply embodied and emotional undertaking that goes beyond individual gratifications of the *here and now*. It is a long-term positioning that calls for continuous aesthetic and ethical recognition. Similar arguments can be found in the gentrification literature, notably in Caulfield's (1994) analyses of early gentrifiers in Toronto, Canada, whose resettling practices in old inner-city neighbourhoods were led by an emancipatory spirit and a search for city spaces unaffected by hegemonic culture (see also Lees, 2000: 392-396). Elective belonging thus represents a structure of feeling in which the typical middle-class desire for autonomy and self-realization materializes through the appropriation of authentic, unexploited places. As such, elective belonging is often sustained through more or less subversive sentiments and counter-hegemonic practices, while at the same time, like a "cultural hypothesis" (Williams, 1977: 132), anticipating broader changes in society.

As Rofe (2003: 2511) argues, "gentrification constitutes a local socio-spatial strategy of identity construction that is increasingly commodified". It is also, one should add, increasingly mediatized. Gentrifying middle-class lifestyles are socially normalized and globally circulated through popular media, where elective belonging is mythologized as a way of being true to one's inner self. One only has to take a quick look at the covers of a magazine like the Swedish *Gård och Torp* (*Estate and Croft*) that targets mostly well-educated home- or second-home-owners interested in the maintenance, preservation and decoration of old countryside houses to see how this works:

"We are living our dream" (No. 11/2015)
"Nothing is impossible – from abandoned house to dream home" (No. 1/2016)
"Bring an old garden back to life!" (No. 2/2016)
"The family rescued the 19ᵗʰ century cottage" (No. 7/2016)

The headlines speak to an audience that actively appropriates and (re)creates space. They are not part of the original local population, but are invited to "rescue" old heritage places from various stages of decay and turn them into expressive elements of their aspirational biographies. While the renovation of old houses is coded as a source of identity creation, the countryside itself is coded as a space of attraction and open-ended "potential". Magazines like this can thus be seen as hegemonic agents of organized self-realization (Honneth, 2004), pre-mediating and exploiting the desire for certain types of places, depicting them as particularly authentic, exciting and/or liveable (see also Chapter 4).

Similar mythologizing discourses are at play in policy documents, marketing material and real-estate prospects that accompany government-led regeneration projects. In a study of the transformation of the Western Harbour area of Malmö, Sweden – sparked off in 2001 by a high-end international housing exhibition called *City of Tomorrow* – I found that the "preferred settlers" were often vividly encoded into the housing concepts (Jansson, 2005). They were not only demographically inscribed but also related to the taste cultures of mobile middle-class lifestyles and the type of privileged urban living that the architects and planners had envisioned for the new neighbourhood. This is an example from one of the housing concepts, called *The Simple House*:

> I imagine as tenants for a terraced house of 65 square meters with two floors: First, Maja, thirty years old. Works full-time at Radio Rix as a media sales-woman, rarely on Saturdays and Sundays. In her free time she socializes with friends and enjoys being close to the centre of the city. Maja is also interested in shares and has bought a lot of Ericsson shares. At the moment she is working on getting a scuba-diving certificate. […] Second, Martin, twenty-nine years old. Works as a pilot for SAS after many years of technical education. […] He likes good food and wine and enjoys cooking together with Maja. Martin's interest in art is considerable and he likes to visit galleries frequently.
>
> *At Home: Stories at a Housing Exhibition, Catalogue 1: 24*

In this case, the promotion of the *new* Western Harbour coincided with the ideology of globalism as it was articulated during the heydays of the IT-driven "new economy" around the turn of the millennium. The formerly industrial city of Malmö, which had undergone a long period of stagnation and identity crisis, was aiming for a new position as a transnational node in network society (together with Copenhagen), and the transformation of the old harbour and shipyard area was designed to symbolize this reorientation. The momentous atmosphere of the new economy was integral to many of the architectural concepts, which articulated contemporary visions of the "smart home" and the "smart city". As several scholars have pointed out, these visions (which are still in full swing) not only speak to a particular form of global competition between cities and regions but also tie these societal aspirations to the aspirations of certain class fractions rather than others. As Allon (2004: 260) notes in an account of the technologized home, if "the figure of the housebound housewife was most certainly the target for the discourses of 'home modernity' which circulated in the mid years of the twentieth century, then it seems that the double-income professional couple, the 'symbolic analysts', 'knowledge workers' and info-elite are now the figures of smart domesticity in the twenty-first century". Similarly, in her analysis of how "smart cities" are visualized in online promotional videos, Rose (forthcoming) concludes that such cities are predominantly encoded as governed by the "smart" management of various data flows and populated by an anonymous crowd of mobile people using various screens.

A broad range of media is thus a part of producing and negotiating the spatial codes of particular places, and this concerns transitory places more than others. But there is also a certain socio-cultural logic to the fact that places that undergo change, where the codes are still relatively open to negotiation, are often designed for and gain attraction among the middle classes who can more readily *recognize themselves* in the very *becoming* of place and place identity. The feeling that one has found something unique and/or is part of creating something new, pertaining not just to a singular place but to the entire cultural-material structure of society, is one that clearly demarcates the aspirational ethos of elective belonging.

However, this type of belonging is not free from conflicts, but is haunted by inescapable contradictions and incompatible ends. For example, what happens to the sense of self-realization and socio-spatial authenticity among early settlers when their neighbourhoods become the subject of more organized forms of transformation and commodification, or when dominant media narratives change? We have already seen that there are tensions between gentrifiers and non-gentrifiers, as well as between different gentrifying groups. What is the cultural-material status of media in these different contexts? To what extent and in what ways are media (techniques, properties and textures) important to elective belonging? What role do media play in the construction of the so-called smart home, and under what conditions is such a networked version of domesticity wanted in any case? As the forthcoming analyses will show, media often make it easier for middle-class subjects to organize their lives in complex ways and achieve recognition for important life choices. At the same time, however, this *mediatized complexity* (as in the term *organized* self-realization) is also what potentially threatens the whole experience of elective belonging.

Contrasts and enclosures in urban gentrification

A recurring theme in my empirical material on middle-class residential mobility is that resettlement is associated with a strong desire to accentuate contrasts within one's current life situation as well as within one's overall life biography. The desire for difference is reflected *both* in the embracing of socio-cultural friction and excitement in everyday life – for instance, by moving to places in transition or to neighbourhoods marked by a social and cultural mix – *and* a willingness to try new things in life, ultimately positioning *oneself as another*. This kind of openness to contrasts and curiosity about the Other has many similarities with the cultural and aesthetic pre-dispositions identified in pioneering research into early gentrifiers such as artists and cultural intermediaries (Zukin, 1982; Caulfield, 1994; Ley, 1996). It also resonates with the privileged cosmopolitan outlook described by Hannerz (1990), Calhoun (2002) and others (see Chapter 5). I would suggest that gentrifying lifestyles can be seen as a form of *domestic cosmopolitanism;* a form of cosmopolitanism that does not require extended global mobility but rather combines an ethically grounded willingness to engage in others with more "banal", and thus largely home-based, forms of cosmopolitan experience (cf. Szerszynski and Urry, 2002; Beck, 2004/2006: 40–42). As Rofe (2003) argues, based on his fieldwork in the cities of Sydney and Newcastle, Australia, gentrification is an inherently *translocal* phenomenon, not just in the sense that in many respects it looks the same in different places but also because it is driven by a class fraction of globally oriented people who seek the rhythm and action of the global at the level of the local, whether we speak of food, design or cultural events.

Geneva 2014

Domestic cosmopolitanism is associated first and foremost with the bustling street-life of inner cities. We can see this in my fieldwork among Scandinavian expats in Geneva. Compared to my other respondents, those who had decided to live in the socially mixed district of Paquis in Geneva, an area notable for its low-wage immigrants, had in common not mainly that they worked for international organizations but that they possessed larger amounts of cultural capital, having a professional background in the cultural sector or living together with cultural workers. As I mentioned at the opening of this chapter, the cosmopolitan orientation is reflected in a desire to "get out of the bubble" and lead a life marked by greater authenticity and heterogeneity, in tune with the egalitarian and emancipatory ideals of international development organizations.

> It's difficult to find housing in Geneva, so at first we stayed at an apartment hotel in Paquis, on the very messiest street. And then we got to know the neighbourhood. We had already heard that there were some cool quarters, a bit boho-chic, with different types of people from all walks of life, different income levels. It's really integrated. There are junkies mixed with sex workers mixed with diplomats mixed with… It's [mixed] in all kinds of ways. It gives the place a vibrant atmosphere. So then I felt I didn't want to live anywhere else, I wanted to live here.
>
> *Linn – technical officer in her thirties*

As argued above, however, residential decisions like this tend to anticipate broader social transformations of neighbourhoods, which implies that gentrifiers often find themselves torn between hegemonic forces and counter-hegemonic ambitions, between a search for difference and socioeconomic mechanisms of spatial commodification and encapsulation (Christensen and Jansson, 2015). Rofe (2003) found during his fieldwork in inner Newcastle, Australia, that gentrifiers who craved cultural diversity and uniqueness also tried to find new ways to maintain their distinctiveness as their neighbourhoods and lifestyles became widely popularized and commodified. Some of his respondents complained that new residents who had bought new, exclusive apartments in the area were "solely interested in purchasing a pre-fabricated identity" or "renting a lifestyle" (ibid.: 2523).

At the same time, signs of gentrification may evoke feelings of scepticism among established communities. In Geneva this tendency is accentuated by the obvious *locational advantages*, in terms of residential access, of expatriate populations (Rérat and Lees, 2010). As Pattaroni and Adlay (2013) suggest, the cohort of international civil servants in Geneva constitutes an "expatriate ethnoscape" (see Appadurai, 1990), which is at once intertwined with local life and detached from it by financial and communicational infrastructure of various kinds. The fact that new-build gentrification, specialized international relocation agencies, online networks like Glocals.com and dedicated Facebook pages make it easier for this group to find housing without having to make a long-term commitment to the city has upset many of the inhabitants of Paquis. Alongside the new housing complexes and so-called "expat bars" that offer English-language services and "cosmopolitan aesthetics", there have also emerged local protest groups (Pattaroni and Adlay, 2013). One of my respondents, Emilia, who lived in Paquis while she held an internship at one of the United Nations (UN) Geneva-based organizations, was also very engaged in the city's sub-cultural scenes. In the interview she describes how the territorial conflicts that emerged around gentrification affected her and other expats who wanted to overcome socio-cultural barriers rather than distance themselves from the city:

Interviewer: It sounds from your experience as though there was a divide between the local and sub-cultural groups and the international environment. Was it possible to combine these worlds?

Emilia: It might have been the case that they [members of local subcultures] didn't respect me very much because I was working for the UN, because that was something they all turned against very strongly: "All those rich bastards coming to the city and making the rents go up and now they want to chase us away from the inner city too because we make noise and the rich people want peace and quiet". It was very much like that.

Gentrifying lifestyles are thus not always successful in their ambition to establish a sense of growing autonomy and positive recognition. A desire to actively resist geo-social segregation, a process that these groups are themselves part of, may lead instead to experiences of cultural cross-pressure.

Accordingly, questions of where to live and which communities to get involved in, within and beyond the local setting, and *by what means*, require a great deal of reflexivity. This also regards the social uses of media. As discussed in Chapter 5, professional networking is difficult to avoid if one works in the UN sector, and these media practices continually bleed into other time-spaces of everyday life. Work-related communication demands constant attention and engagement. At the same time, it is obvious that the cosmopolitan striving to break out of the expat bubble influences how media are used in the context of private relationships. While all respondents in the Geneva study stressed the importance of social media for staying in touch with friends and family in *other parts of the world*, such media were rarely used for locally oriented purposes. Even though there are many online communities and applications designed to bring together international people in Geneva, none of my respondents had actually used these tools for social purposes. Rather, in line with the above-mentioned spatial privilege of middle-class expats, geo-social media had been used for practical purposes, especially for *finding accommodation*. Among those living in the Paquis district and oriented towards more open-ended "street-based" lifestyles, this pattern of media use reflects a principal ambiguity. While specially designed media may be used for generating a short-cut into the Geneva housing market, and while social media may be embraced in order to maintain ties with international friends, there is a stated desire not to contribute to further geo-social encapsulation. Geo-social "expat apps", like the ones studied in Polson's (2016) work, were not valued by the UN professionals I interviewed in Geneva.

Camilla, who works on a five-year deal with a UN organization (following a strict rotation policy), represents an interesting case of "in-betweenness". Previously she and her partner lived in the inner city, in an apartment they found through an online community, but after becoming parents they decided to move to a quieter neighbourhood with more space, though still *not a typical expat area*. Camilla explains that she is afraid of becoming "too comfortable" and does not want to be associated with some of the UN people with "their houses and swimming pools and security and fat salaries". Similarly, she has never been interested in mediated social communities.

> Since I'm not using these communities myself it's difficult to know exactly what types of people there are there. Probably quite a few that are single and don't have a partner or family. Perhaps more outgoing people who enjoy socializing with lots of others. I guess we're not really like that. We entertain less and have deeper friendships rather than a very loose network. And then it's also the fact that I feel that I'm so incredibly fed up with that kind of conversation, when you are at a party and people ask what work you do, how long you've been here and so forth. […] Here we have a little community with the neighbours, with a gardening cooperative and so forth. Otherwise we don't meet many locals. Previously [when we were living in the inner city] there was a friend of mine who had a boyfriend who was from Geneva, as well as other friends of friends. There was a German girl who was very good at looking up what was going on in the city; you could just follow her and find things to do. Through her I also got to know people in the cultural scene who were working with art and music and so on. But then she moved back to Germany and we had children and now we don't have much time.
>
> *Camilla – programme communicator in her thirties*

Camilla's story is important for the present study in two ways. First, it shows that elective belonging is an elastic condition; most of the time there are compromises involved in residential choices, especially during the family phase. Of course, it is a spatial privilege to be able to make residential choices like the one Camilla describes, but this does not mean that there are always strong feelings of local attachment involved.

Second, Camilla's story problematizes Polson's (2016) arguments concerning the role of geo-social media in the formation of global middle-class fractions. Having studied this phenomenon in Paris, Singapore and Bangalore, one of Polson's key findings is that the use of geo-social media tends to reproduce the stratified order of expatriate lifestyles. While the most prestigious groups in her study were using dedicated platforms and gathered at exclusive venues that required a special invitation, those who were less established, and often at an earlier stage of their careers, sought to make connections in a more open-ended way, using less exclusive networking tools and therefore interacting with a more diverse group of people. What is missing from Polson's study, however, is the *non-use* of geo-social media (since she was only interviewing people who actually used these platforms) and what such an orientation means in terms of status and distinction. My point is that the orientation towards cultural capital implies a relatively high degree of social reflexivity in relation to (and a relative avoidance of) connective media, corresponding to the questioning of the geo-social encapsulation of expatriate communities (Jansson, 2011; see also Chapter 4). The progressive, gentrifying lifestyles I have described here illuminate how elective belonging goes hand in hand with *elective media use*, in the sense that different media resources are used and appropriated in different ways in order to maximize the possibility of making residential choices that can sustain culturally distinctive lifestyles. Connective media are embraced as a means to get easy access to social unpredictability and urban authenticity (*mediatization as spatial privilege*) while at the same time kept at a certain distance in order not to disturb that very same sense of everyday cosmopolitanism (*mediatization as cultural restraint*). Media are thus appropriated as tools for locational advantages that these gentrifying cosmopolitan subjects are themselves critical of.

Malmö 2003

The tension between social experimentation (contrasts) and homogenization (enclosures) is not restricted to progressive inner-city lifestyles. Middle-class fractions who are less experimental, who settle in provincial areas or buy new apartments in regenerated post-industrial areas (and thus represent those who can afford to buy "pre-fabricated lifestyles") are affected by similar cross-pressures. In the above-mentioned case of the Western Harbour in Malmö, Sweden, the whole transformation of industrial wasteland, including the housing exhibition, was initially celebrated in local as well as national and international media as an example of smart, sustainable urbanism and social change. Many of those who moved into the apartments in the former exhibition area not only invested great amounts of money, they also took a deliberate risk in the sense that the future of the area at that point only existed in planning documents and on drawing boards. The interviews I conducted among early settlers show that the desire for change, difference and open-ended spatial production was an essential part of the decision to move. What was neither desired nor fully expected, however, was the negative recoding of the neighbourhood, led by dominant news media, that escalated after various scandals related to the housing exhibition (such as bad production quality and corruption) and was fuelled by the global downturn of the "new economy". The housing exhibition went bankrupt in early September 2001 and was followed by a period of criticism in local and national news media in which the Western Harbour was depicted as a failure and a "rich man's ghetto" (Jansson, 2005). The following extract, taken from an interview with Roy, a middle-aged restaurant owner in the Western Harbour area, gives a vivid illustration of how the middle-class ethos of elective belonging and self-realization collided with the public recoding of place:

> This area, when you walk around here, you don't know for sure if you are in Nice or somewhere else. But at the same time this place has its own character. Look at that pier out there [pointing out of the window], it looks like it could have been in Brittany! To me, the sea is very important, and to feel that one is part of building up something new, that is what fascinates me! I think it was completely wrong when they previously used this area only for industrial purposes. Completely wasted. When I walk these streets I get the feeling that something new and exciting is being built, and I don't know if there are many places like this, where those who live there have such strong feelings for their neighbourhood. […] It's a very special place that even gets attention abroad. Every week there are people visiting from other countries, entire buses as well as individual visitors who have also visited Copenhagen. They come in here and talk. I've almost become like a guide for this neighbourhood, and I love to defend it one hundred per cent. I really love walking around here. […] But the entire exhibition received nothing but savaging, whatever the subject was… Just look at that pier out there: it's unique, so beautiful! That's something to write about!

Roy's outlook corresponds to the image of gentrifiers as a globally oriented class-fraction at the forefront of social change. Before starting his restaurant Roy had seen much of the world during his extended travels and was particularly inspired by the Italian Slow Food movement. He was attracted to the Western Harbour mainly because of its environmentalist profile and its strong symbolic ties to the rest of Europe and the world (situated with a panoramic view of the new Oresund Bridge between Sweden and Denmark). The Western Harbour provided a *space of potential*, a *different place* where he could test his progressive ideas around sustainable dining and reach the right clientele for such a restaurant. While his estimation of the market was correct, and the restaurant became widely known, for a long period he had to cope with the dissonances of negative media coverage and his own appreciation of the Western Harbour. Mediated stigmatization is a phenomenon that otherwise mainly affects socioeconomically underprivileged neighbourhoods, in Malmö as well as elsewhere (e.g., Stjernborg et al., 2015).

Roy shares this ambiguous experience with many others who moved to the Western Harbour during the first years of the transformation. Most of the people I interviewed in 2003 described how the move had coincided with, and in itself symbolized, a new direction in their life biographies. The settling in an entirely new neighbourhood, a place without any residential history but a relatively open-ended script, implied an opportunity for self-reflection, even self-transformation, and was as such influenced by many different sources. Media images came to play an extraordinary role because they influenced the attitudes of people who themselves had not been to the Western Harbour or even to Malmö.

Anna is a middle-aged dentist who moved to her apartment after a divorce when she felt a strong need to start anew. Her experience of ambiguity and ontological insecurity was reinforced by the fact that the new district was often criticized in news media for being a socioeconomic failure and a (segre)gated space. At the same time it figured in design magazines and fashionable advertisements as a space for tasteful post-industrial lifestyles. Anna's choice of residence thus caused a lot of curiosity and mixed comments among her friends in other parts of Sweden, which reinforced her sense of biographical reorientation. Her whole sense of identity was mediatized, fluctuating between self-recognition and stigmatization.

Anna: I subscribe to *Residence* magazine and they often have adverts from real estate agencies and I can see that Western Harbour often figures there along with, well, what you may call exclusive districts in other parts of the country, Stockholm and Gothenburg and other places. So I can see that it's now in company of a certain kind… that they put it in a very exclusive context, that's how it feels.

Interviewer: Do you identify more strongly with the area because of such ads or are they disturbing?

Anna: I'm a little bit, like, I don't know which leg to stand on because I want to live in a neighbourhood that feels… not exclusive but it should be neat and clean in some way, but not boring. But I don't want the place to be labelled as exclusive or snobbish, because I like the integration and so forth. But I want it to be nice and clean and good quality and I want to identify with good architecture and good materials. I like it when it suits me, and it may not be the taste of others, but I like to surround myself with things that I find beautiful. I can pay some extra for that but I don't want it to be flashy. Can you see what I mean?

Anna's experience provides an interesting narrative of the complex recoding of the Western Harbour and how it affected its inhabitants. It also encapsulates the more general insecurities that characterize gentrifying middle-class lifestyles. Elective belonging, as a structure of feeling, should not be mistaken for a straightforward engagement with spatial production or as the simple fulfilment of individual life-goals. Rather, this type of middle-class investment in place is accompanied by a great deal of reflexivity and, as I discussed in Chapter 6, *hesitation*. While Anna says that she "likes the integration" and thus indicates that she wants to embrace cosmopolitan values of hospitality, openness and social mix (Dikeç, 2002), in contrast to the exclusivity and "snobbishness" she identifies in residential magazines and the like, she is not willing to compromise her own sense of security and quality of life (understood also in material terms). This means that she is aware of the process of geo-social encapsulation she is herself part of, and in different ways reminded about by the media, but is still not willing to fully identify herself with it. This type of hegemonic self-reflexivity resonates with the underlying anxieties of middle-class habitus, which has been analysed in several previous studies on gentrification and middle-class residential trajectories (e.g., Zukin, 2008; Abrams et al., 2013; Benson, 2014).

Given that my study in the Western Harbour was carried out in the early 2000s one can only speculate as to how connective media and mobile applications might have influenced the spatial recoding processes. It is reasonable to assume that a more multi-faceted and interactive technological order, including for example Facebook groups and other online communities, would have increased the opportunities for new residents to gain mutual recognition and maintain an even stronger sense of spatial attachment. Even at the time of my project most of the respondents mentioned the importance of an independent news website called *Västra hamnen* (*The Western Harbour*), which was run (and still is) by one of the early settlers (a photographer and computer programmer). News and images related to the neighbourhood were gathered and posted on the website every two days. The website was basically run as a hobby project, but gradually came to play a key role in establishing a local community, even functioning as a phantasmagorical resource for prospective inhabitants who were able to get a taste of "what it would be like to live there" before actually moving there.

The popularity of the Western Harbour website underscores that locally based "grass root" communication can contribute to solving symbolic ambiguities and threats. What the website did, above all, was to *reinforce a sense of elective belonging* among its followers. When I interviewed the keeper of the website in 2003 he claimed that it would not have been possible to do the same thing in any other neighbourhood in Malmö because nowhere else was there as much to report on and nowhere else were the local inhabitants as engaged as they were in the Western Harbour. Online community media thus became a way for the settlers to reclaim the uniqueness and authenticity of the area and to accentuate their sense of changing the future of Malmö. As such, the Western Harbour website invited people "to *feel* their own place in current events" and thus constituted a source of what Papacharissi (2015: 4, italics in original) calls "affective publics". While this mobilizing role of digital media has since been further strengthened through social media platforms, the Western Harbour website also anticipated the kind of virtual bubbles that to an increasing extent operate as shields against external influences and criticism in today's cultural landscape. Mediated mobilization and encapsulation processes tend to go hand in hand.

Topophilia and exitability in provincial gentrification

At the core of elective belonging is the *emotional attraction* to a particular place and the desire to *create* place in a way that fits one's personal identity. Elective belonging thus implies the coming together of cultural pre-dispositions (taste) and embodied practices that connect identity to material geographies. However, the *electiveness* of elective belonging also implies that the bonds to a particular place are negotiable. There are no inherited roots, family relationships or traditions that prescribe a certain residential trajectory. As we saw in the interview with Roy, for example, elective belonging has more in common with a modern love relationship: a sense of having found the right place and a commitment to build and entertain a strong and lasting relationship with that place. I call this relationship *creative topophilia*. Human geographer Yi-Fu Tuan (1974) speaks about *topophilia*, literally meaning "love of place", to describe why people become attached to a place. According to Tuan (ibid.: 93), topophilia is a composite condition that "can be defined broadly to include all of the human being's affective ties with the material environment". These ties may be primarily aesthetic, tactile or related to deeper attachments, memories and a sense of home. As Savage et al. (2005) argue, elective belonging principally rests on the former elements rather than on historical roots. In contrast to "nostalgia", elective belonging is a feeling of attachment that is *produced* through more open-ended (but still socio-culturally structured) encounters and grows stronger through cultural and material *work* with a place. Thus the term *creative* topophilia.

In my material on mediatization and middle-class trajectories there are many examples of creative topophilia. They are related to inner-city lifestyles as well as to new-build gentrification in post-industrial environments. More than anywhere else, however, I have found expressions of this feeling among privileged counter-urbanizers in provincial Sweden. In this context it is not just the love of place, topophilia, that is vividly, even poetically, articulated. It is where I have also found particularly strong evidence of the middle-class orientation towards creativity and self-realization, meaning that the establishing of a home-place, precisely because of its elective nature, remains an ongoing *project* rather than a final destination. Middle-class dwellers in the countryside are torn between a desire to discover and create places that are really "places of their own", sometimes characterized by their very remoteness, and a felt need to maintain key social networks and connections to other places, ultimately for the sake of keeping the exits open. Under such conditions, as I have argued elsewhere (Christensen and Jansson, 2015: Ch. 5), the role of mediated connectivity is bound to be particularly dynamic, reinforcing both topophilia and exitability and often leading to contradictory experiences.

My fieldwork was carried out in Arvika, a municipality of 26,000 inhabitants in the western part of Sweden, where approximately 14,000 people live in the town of Arvika itself. The municipality can be considered a rather typical "space of potential" in that processes of industrial and, before that, agricultural stagnation are paralleled by efforts to sustain new forms of socioeconomic development tied to more service-oriented businesses, including tourism and culture, in order to attract new inhabitants and visitors. During a four-year period (2010–13) I interviewed, together with my co-researchers, a social mix of people in Arvika and its surroundings, including separate samples of Dutch immigrants in the countryside, first- and second-home owners in a rural community, and artists and craftspeople in Arvika and the vicinity (see Chapter 1). The latter two groups consisted mainly of return migrants ("homecomers") and in-migrants ("newcomers"), who were attracted to Arvika because of the region's scenic nature and resources for outdoor interests, combined with its rich cultural heritage and relatively lively cultural scene. In the municipality there are wilderness reserves and hiking areas as well as a university campus for music education, culturally oriented colleges for adult education, a well-renowned art museum, several galleries, craft shops, and venues for music and theatre performances. As one of the interviewed artists, new to Arvika, commented, "We knew that Arvika was a bit more exciting than the average backwater, and that many had stayed or moved here". Since my respondents were rarely driven by strong ideological agendas, they can be positioned at the "gentrifying" rather than the "radical" end of the spectrum of counter-urbanizers (see Halfacree, 2007, 2010).

Second-home ownership

The electiveness of country living is for obvious reasons most apparent among second-home owners, who can afford to maintain a relatively romantic view of what it is "to be rural". At the same time, these cases provide important insights into the factors at play in gentrifying counter-urbanization, including mediatization. Lili and her husband are both academics, and at the time of the interview they had owned their old countryside house for about ten years. During this decade they had spent not only vacations and weekends there, but also longer periods, including one full-year stay when they were partly on parental leave and partly working. They had also spent several years abroad, working and studying in different countries. It took them about a year to find their home in the country. They searched in many different regions for the right place – "a place to work with", as Lili put it – and eventually found an original house to renovate, perfectly located close to a lake in a rather remote area without any close neighbours. While they immediately felt that this was "their place", they also wanted to restore the house and the surrounding property in a way that was true to its history. This balancing between, on the one hand, creating and restoring things and, on the other, preserving the feeling of authenticity, also resonates in how Lili describes the couple's attitude towards media. According to Lili, the material presence, or remnants, of older media infrastructure provided important links to the history of the place, as well as to a bygone era of a slower pace of life and closer connections to the natural landscape.

> It's a bit like the fact that the name of this place is actually on the map and the whole media historical connections of this place, and the amusing idea that this place was established during a very primitive time in the past when the land was cultivated and stones removed and was then reached by modernity

and new technological achievements and became part of that network. We can still maintain some parts of it. Keeping this phone number, for example, amuses us. I feel a great joy that we have a phone number connected to this very place and that we can see the cable that runs outside the window here. We know that this is the connection out. It would not have been the same thing if we only had mobile coverage here. And we can still even see the remnants of cable radio. So… absolutely, there is something romantic about it too.

Lili – researcher in her mid-thirties

The presence of old media, even such material traces as the sockets of cable radio (a system that was established in Sweden during the 1940s and in operation until the 1970s, providing radio through the telephone network to the remoter areas of the country), can therefore contribute to a sense of spatial authenticity and attachment. The discovery of historical layers of infrastructure and the preservation of place-specific markers, like the authentic phone number, form part of the composite appropriation of a place upon which deeper experiences of topophilia can be built. The fact that second-home owners do not have to adapt their places to contemporary standards of living also implies that these places can be designed as counterweights to mediatization processes. Second-home ownership becomes a privileged way of negotiating media dependences, claiming greater freedom and a sense of autonomy in relation to mediatization processes.

Lili: I remember with great pleasure the times we lived here or stayed for longer periods when we had our daily walk up to the mailbox as a nice little ritual. It contributed to the "for real feeling" we talked about and requires very physical manifestations, like connecting the house to the cable or setting out the mailbox, which somehow brings along a whole media infrastructure. To us the feeling of being a little "outcast" has been part of the charm and the fascination that "out here", in the remotest place, at the end of the cable, that's where we are. The time we lived here was a fascinating experience and we used a telephone modem to go online and I could even do a university course when we lived out here. This was before Facebook, Spotify and that type of constant online streaming.

Interviewer: But is there any type of media technology you miss when you are here today?

Lili: No, but I'm also a rather special person! [laughter] It's actually been ten months since I lost my mobile somewhere around here and I still haven't replaced it. I think that this is a phase of life, a period of downscaling in terms of communication. So I miss absolutely nothing. I'm just happy not to have social media. Out here we're listening to the radio a lot and that's a kind of… slow medium. You can really feel that.

What is particularly interesting to note here is that mediated connectivity – whether we speak of the postal system or mobile networks – actually reinforces the feeling of remoteness. The feeling of being "the last house on the line", in turn, generates a sense of independence, being located on the very margins of media society but still having the opportunity to be included. This is to say that the marginal geographical location, where not only the creation of place but also the maintenance of connectivity demands work, feeds into the emotional structures of elective belonging. Owning a second home where one can stay for longer and shorter periods means that one has access to a space where one can regain *control over connectivity*. As such, it is a typical space of creative topophilia where different options can be kept open – disconnection versus reconnection, old versus new media, and so forth – and negotiated in relation to practical, emotional and existential circumstances.

Basically, this counter-hegemonic spirit can be identified in most of the interviews with counter-urbanizers. However, as soon as leisure-oriented ways of life evolve into something more permanent, connectivity becomes a more delicate issue to handle.

Lifestyle migration

In my research there are several examples of urban people who have settled permanently in their summer residences in Arvika. All of the Dutch immigrants, for example, had familiarized themselves and "fallen in love" with the region as tourists and on longer vacations and had eventually decided to move there. These life-transforming decisions are accompanied by very engaging stories of creative topophilia. For instance, a middle-aged Dutch woman described how she cried with happiness on the ferry when she saw the Swedish coastline and how she had never before felt as much "at home" as after her settlement in the countryside of Arvika. Throughout these life stories one can also see how the meanings and practical handling of media technologies and infrastructure are negotiated. Moving to the country *for real* clearly implies that one can no longer fully disconnect. Separation as such is not a desired state. Nevertheless, most counter-urbanizers still agree that the countryside location works as a sanctuary from information pressures and other forms of external stress, in keeping with the general mythology of rural living (Jansson, 2013b). They describe their country lifestyles as a way of enhancing their level of individual autonomy, gaining the opportunity to retreat to *a place of one's own* without giving up other connections.

> I didn't want to live too far out in the country. We live in the country, but still, it's forty-five minutes to Karlstad or twenty minutes to Arvika. So, that's one of the reasons, because of the unspoilt nature… And you know, with the Internet it doesn't really matter where you are. And I have to say, I think it's very good that you do have everything in Sweden, what you need, even when you are living out in the country. […] I mean, for me, it's important that you can easily leave if you want to go and see a lot of people. But it's so nice to go back to the countryside again.
>
> *Mirijam – a middle-aged entrepreneur and Dutch immigrant*

Mirijam's story is interesting, and typical, in the sense that both the peacefulness of the countryside *and* the extended connectivity are stressed as important. Mirijam had lived in several big cities in the world before she and her husband decided to look for something completely different. The main attractions of the countryside and the Arvika region were the access to wild nature and the opportunity to get *more space* than they could afford in the Netherlands. Still, the move would not have been possible if there had not been good media infrastructure. Mirijam and her husband run a company oriented towards healthcare, the same company they ran when living in the Netherlands. The main difference is that they have become even more dependent on well-functioning media systems since they moved to Sweden in order to maintain their customer networks. The company is physically located in their home, a rather big heritage house with several additional buildings, including barns and stables. They also keep animals on their property and run a small flea market in the oldest building. Altogether, they have put in a lot of hard work and created very strong attachments to their new home-place, including making great efforts to maintain connectivity.

> We have a huge professional network. Internet of course, but mainly through different networks that we have. And our sales and marketing office in the Netherlands that takes care of that now. [...] Actually we are always busy here. We have these animals, and everything needs to be done, but mainly what I'm doing is the treatment, doing research, and working on that, so there is no vacation. We are always working, so... In three weeks' time we will have our first week off in what? Twelve years. So we never have a day off, we always are working, there are always people here. It's fun, it's fun. Our work is our hobby. Or our hobby is our work.

The fact that Mirijam and her husband can live and work at the same place, thanks in large part to media technologies, means that their home-place becomes a means of linking different elements of their life biographies into a coherent whole. Or, seen from the opposite perspective, the bringing together of different spheres of life into one place implies that this place is invested with existential authenticity and a sense of biographical resonance. These observations correspond with previous research on *lifestyle migration* (e.g., Benson and O'Reilly, 2009; Benson, 2016), showing that people who move to more or less phantasmagorical places, largely driven by their touristic or leisurely preferences, must work hard in order to maintain a sense of biographical continuity and not feel cut off from their roots and pre-established social networks. At the same time, middle-class residential (and in this case professional and entrepreneurial) trajectories contribute to the gradual transformation of habitus (Benson, 2014).

What is obvious from Mirijam's story, as well as from other interviews with counter-urbanizers, is the multi-layered importance of mediated connectivity. It is important on three principal levels. First, connectivity involves a strong *symbolic* component, as we saw in the case of second-home ownership. Being able to connect, even though such connections are not used regularly, is a potent symbol of social integration and often addressed as a place-political concern among rural dwellers (Jansson, 2010b; Jansson and Andersson, 2012). Second, connectivity has an *everyday practical* significance for country lifestyles when it comes to the maintenance of leisure interests, education, work and various social relationships. Media also contribute to the blurring of boundaries between these social spheres. Third, connectivity is important for mobility from a long-term perspective, ultimately working as a pre-condition for *exitability*. The fact that Mirijam and her husband actively maintain their professional networks, as well as other social relationships, not only secures the operations of their company and the continuity of their biographies, it also makes it possible for them to actually change their minds and move back to the Netherlands or to some other place if they wish. Exitability is the ultimate expression of the privilege that comes with elective belonging. If one gets stuck in place, or is unwillingly displaced, belonging can no longer be seen as elective.

Cultural entrepreneurship

The significance of exitability is also obvious among the artists and cultural entrepreneurs we interviewed in Arvika. What they had in common was that they were well established within their fields, possessing significant amounts of cultural capital (including a Masters of Fine Arts or other university degree) and large professional networks on both national and international levels. Furthermore, their trajectories were characterized by the typical push-and-pull factors of counter-urbanization. As well as environmental advantages like proximity to nature, shorter distances of travel and a more harmonious pace of life compared to the city, they emphasized the lower costs of living in Arvika and the opportunities to buy or rent large spaces where a dwelling and studio work could be combined. Settling down in Arvika was thus an attempt to combine autonomous artistic work with a greater overall quality of life without losing one's position within the field.

Keeping one's professional position, including valuable social networks, was seen as the key to exitability. However, moving to a small town also means that the individual encounters a particular cultural community with its own unwritten rules and regulations; that is, a place-specific *doxa*. A key asset for entering such a community is to possess a locally recognized type of social capital, understood in the Bourdieusian sense as "the sum of the resources, actual or virtual, that accrue to an individual or a group by virtue of possessing a durable network of more or less institutionalized relationships of mutual acquaintance and recognition" (Bourdieu and Wacquant, 1992: 119). Accordingly, while many of the homecomers felt that their re-integration in Arvika had been a smooth process, even gaining some attention in local news media, the respondents who were not from Arvika and who had made a more uncertain investment in their new hometown described the opposite.

When I phoned the local newspaper, I… […] Well, I explained what I did and why I wanted them to come. Said I was new to Arvika and invited them

> to an exhibition. [...] When I finished my spiel he said, Well, who are you? [...] He was fishing for some kind of connection to Värmland or relatives etc. And then there was no interest. So they never turned up.
>
> *Claes – visual artist and set designer in his forties*

> I think it's pretty parochial. Cultural life is very closed. [...] But, by God, it's not easy to be a cultural artist here, especially not as a new resident. This is evident, socially and professionally.
>
> *Greta – middle-aged visual-arts teacher and entrepreneur*

Newcomers to the local environment described that it did not help much that they brought with them international degrees, large networks and impressive records of exhibitions, grants and invitations to national and/or international institutions. These resources were sometimes even met with suspicion, undermining their chances of achieving local recognition for their work. As a consequence, it became even more important for the newcomers to stay connected to the epicentres of their artistic field; that is, to maintain and actively manage their networks as a kind of professional insurance – as a key to the exit. Newcomers thus found themselves in the cross-hairs between local doxa with its insistence on obedience to inherited orders of recognition and the dominant, (inter)nationally defined, logics of the artistic field with its tendency to ignore artists in provincial areas. As revealed in the above extracts, the frustrations and feelings of being "out of place" were expressed in a vocabulary that even reproduced the divide between centre and periphery they had set out to overcome.

Again, this shows how symbolic power geometries are inscribed in the physical landscape, making places like Arvika seem and feel more or less "remote". The interviews also show that the composite force of mediatization is experienced in different ways depending on *where* social actors, in this case counter-urbanizing artists, are socially and geographically *located* and *anchored*. While the artists generally pointed to the importance of staying tuned to activities going on in other places, notably in Stockholm and other bigger cities, this was emphasized especially by those without any previous connections to Arvika. As one respondent put it, he "preferred to engage in face-to-face meetings rather than reading the minutes". However, if geographical and economic factors make it more difficult to travel, set up face-to-face meetings or participate in cultural events, strategic forms of online presence are bound to gain prominence.

> I use Facebook a great deal in my profession. To me Facebook is a means of contact with a professional network. My network then grows, everyone you have worked with, all your fellow course participants who are active around the world. It's an enormous network!
>
> *Lovisa – visual artist in her forties*

While moving to the countryside is fundamentally described in terms of elective belonging, a way of finding a more harmonious and place-based way of life, the other side of the coin seems to be that a peripheral location – geographically and socio-culturally – accentuates the indispensability of various media resources and skills, especially among artists who want to entertain the possibility of restoring their trajectories and returning to more central locations. Ultimately, this means that *the indispensability of media is reinforced by the artists' desire for artistic autonomy*. Growing media dependence, in turn, implies that more time and effort must be invested in various forms of strategic management, which ultimately disrupts the conditions for artistic authenticity as well as elective belonging.

In short, the example of counter-urbanizing artists, together with the previous cases, shows how the combination of geographical and socio-cultural mobility tends to accentuate the dialectic of mediatization. In the concluding section I will summarize these findings and discuss how they can help us to theorize the linkages between mediatization and gentrification.

Conclusion: Mediatization as a mobility regime

In this chapter I have used gentrification as the analytical lens for studying the dialectic of mediatization. Gentrification refers to the displacement of less privileged social groups and their ways of life from certain districts and neighbourhoods as a consequence of social upgrading. It can thus be seen as the main spatial consequence of middle-class residential mobility, whereby the cultural *and* material structures of place are gradually altered ("commodified", "surveilled", "globalized", and so forth). However, gentrification is not simply the outcome of individual preferences and choices. It is also shaped by broader economic and political forces that strive to maximize spatial productivity through various forms of regeneration projects and the promotion of new consumer goods and services. There are thus two stories to tell about gentrification. One is about the "emancipatory city" (Lees, 2000), the search for social contrasts and transformative self-realization through spatial creativity by the relatively well-educated but not necessarily affluent middle classes. It is also about the production of *hospitable places* of social mix and openness, where topophilia and cosmopolitan outlooks are not incompatible (e.g., Dikeç, 2002). But the other story is about gentrification as the spatial manifestation of capitalist hegemony embodied in converging, and thus segregating, middle-class residential trajectories. If we bring these stories together, which I think we should, we arrive at an understanding of gentrification that comes close to Honneth's (2004) view of organized self-realization: gentrification constituting a structural, but also contested, mediator of normalized social expectations. We can even think of gentrification as *the spatial logic of organized self-realization*.

As I argued in Chapter 4, the ideological regime of organized self-realization contributes to the legitimization and cultural moulding of mediatization processes, making certain forms of media particularly important for the creation of self-identity. Above all, new forms of connective media provide indispensable channels for the management of middle-class-biased orders of recognition. The relationship between gentrification and mediatization could be described in a similar way. As my analyses have shown, the realization of gentrifying trajectories, whether we speak of counter-urbanization or inner-city lifestyles, demands social recognition. Not only are the cultural and aesthetic attributes of such trajectories widely circulated in popular media and scripted into the very transformation of places, as we saw, for example, in the analysis of urban new-build gentrification. Specialized media infrastructure may also provide locational advantages for particular gentrifying groups, helping them to find housing that fits their taste, as we saw in the case of expatriates in Geneva. Furthermore, gentrifying subjects need various forms of mediated connectivity and recognition to feel secure in their new residential settings, enabling them to achieve a sense of biographical continuity *and* self-realization.

In other words, an important driver of mediatization is the desire to reinforce the experience of *elective belonging*. This is also why I have argued in this chapter that we should use elective belonging as the analytical bridge between gentrification and mediatization. Elective belonging is the particular structure of feeling, and an experiential register of mobile lives (the overarching structure of feeling), that underpins gentrifying life trajectories and necessitates certain forms of mediated recognition.

On a societal level elective belonging is also subject to far-reaching measures of organization and exploitation. As previous research has shown, gentrification invokes a hegemonic transition from relatively experimental and ethically framed ways of living into more encapsulated, commodified and homogenized ones. This transition is bound to generate ambiguous experiences among gentrifying groups, whose ethical outlooks are typically marked by cosmopolitan ideals of inclusion and hospitality and whose strivings for cultural distinction require a certain degree of uniqueness and authenticity. As my analyses have shown, middle-class groups tend to be reflexively aware of these complexities and of their own privileged positions compared to other, potentially displaced, groups. It is also obvious that media play an important role in these processes, not just in raising awareness of local conditions, but also in the very handling of ambiguity. In my research there are examples of how the avoidance of socially segregating media forms may correspond to broader place-political ambitions, what I have termed domestic cosmopolitanism. There are also examples of the opposite, of how locally oriented media practices contribute to spatial attachment and stronger identification with the locality and thus work as a potential shield against social criticism.

Against this background I want to highlight the role of mediatization as an increasingly important *mobility regime*. A "regime" is here understood in terms of the cultural materialist perspective as a *structure of ordinary opportunities and restraints* that frames and conditions a particular set of activities – in this case mobility (see Chapter 2). More concretely, it means that various resources of mediation – technics, properties and textures – become increasingly indispensable for the managing of mobilities in general, and mobile lives in particular. Thinking of the relationship between mobility and mediatization in this way brings us back to the basic understanding of mediatization as a dialectical force that expands through the normalization of ordinary culture. As such, the normalization of mobility as a key ingredient of middle-class lifestyles contributes to the power of mediatization.

Recognizing the role of mediatization as a mobility regime also raises questions of autonomy and power. Gentrification, which ultimately constitutes a *public display of the uneven distribution of mobility as a social resource*, provides a particularly clear illumination of this. While some groups have little opportunity to decide if and when they want to move, others are in control of their mobilities and can make creative decisions concerning where to live, where to work, where to shop and where to travel on vacation. These inequalities are all sharply reflected in the gentrification of neighbourhoods. Elective belonging is a sign of privilege and power not just in the sense that the upwardly mobile middle classes are relatively free to settle in places that contribute to their sense of self-realization, it also marks their *power to leave* if they no longer feel that they belong to the place. In my analyses I was able to identify this privileged condition, *exitability*, across different contexts. It was clearly expressed among the professional expats in Geneva, whose locational advantages contrasted with the (lack of) mobility resources among other migrant groups in the inner city. It was also obvious among counter-urbanizers, whose life stories were starkly different from many other stories about rural populations (typically forced to move from stagnant areas in order to find jobs or stuck in a place because they are unable to sell their property; or, in areas of high leisure-oriented exploitation and gentrification, marginalized on the real estate market because of escalating prices).

Among counter-urbanizers it is also obvious that the ability to control mobility, a source of power that has also been called *motility* (Kaufmann, 2002), is dependent on media access and use. On the one hand, moving to the countryside is often promoted as a deliberate choice to get away from the escalating information flows and connectivity demands of urban lifestyles. The decision to settle in a provincial area can thus in itself be seen as marker of privilege and independence in relation to mediatization processes. On the other hand, disconnection is not seen as a sustainable alternative, especially because it would abolish future possibilities to move or gain other socio-cultural benefits. For example, my research points to the challenges facing in-migrating artists and cultural entrepreneurs in provincial areas who need to gain local recognition and anchor themselves in a new place while at the same time maintaining their connections to a broader cultural field of production.

Dealing with mediatization and dealing with mobility thus becomes an interwoven and sometimes intricate undertaking that does not necessarily, in spite of the relative privilege attached to such negotiations, lead to any deeper sense of autonomy and liberation. In short, what I have attempted to show in this chapter, extending the analyses of Chapters 5 and 6, is that mediatization constitutes a mobility regime that normalizes various new forms of reflexivity and socio-cultural hesitation, especially within the middle classes, and thus not only reproduces but also problematizes the construction of power and privilege.

8

RETHINKING MEDIATIZATION, MOBILITY AND SOCIAL POWER

This book is a response to a number of challenges surrounding the meta-process and research field called "mediatization". A first challenge concerns conceptual clarity. Meta-processual expressions like globalization, individualization and mediatization are important tools for grasping overarching directions of social and cultural change, but can easily turn into open-ended realms of disparate ideas and discussions. While we need multiple perspectives for mapping such complex research areas, it is often difficult to establish clear demarcations and a broadly shared view of what it is we are trying to understand and explicate (see Ekström et al., 2016). My response to this challenge has been to advance a *critical approach* to mediatization, based primarily on the cultural materialist theories of Williams and Bourdieu (see Chapter 2). From a cultural materialist perspective mediatization is defined as a hegemonic transformation of *ordinary culture*. As such, it "begins" with the formation of media as *cultural form* (Williams, 1974) and the normalization of media within (communicational) *doxa* (Bourdieu, 1972/1977, 1997/2000).

Another challenge concerns the question of whether mediatization is "good" or "bad", whether people in general are socially empowered by it or not. A key aim of the critical approach is to highlight the *dialectical* nature of mediatization, the fact that growing possibilities for human agency typically come with various forms of media dependence. Accordingly, the workings and consequences of mediatization can only be addressed through empirical studies. In order to achieve a clear understanding of the inner tensions and complexities of mediatization, I have in this book focused on *mobile lives*; that is, mobile middle-class trajectories and lifestyles in which levels of both media access and the expectation of self-realization are high. Through situated analyses in three mobile contexts – elite cosmopolitanism, post-tourism and gentrification (Chapters 5–7) – I have shown how mediatization evokes experiences of ambiguity and a corresponding need to establish morally and culturally recognized ways of managing media-related opportunities and restraints in everyday life. Our academic awareness of routinized and context-specific modes of "media management" such as these makes it possible to formulate an *immanent critique* of mediatization as a social force.

Along these lines, in this concluding chapter I want to bring together my ideas and findings to form a comprehensive picture of the relationship between mediatization, mobile lives and social power. The aim is to transcend the context-specific ways in which mediatization works and is perceived (see Chapters 5–7) in order to formulate more general points about how mediatization is intertwined with – moulding and being moulded by – the socio-cultural struggle for autonomy and recognition in a mobile, ultimately global, society. I begin the chapter with a summary of the cultural materialist approach that highlights the relevance of a *critical bottom-up perspective* for understanding mediatization. In the second part of the chapter I extend the dialectical aspect of my approach in order to unpack the relationship between intensified mediatization and various forms of cultural resistance and critique. Envisioning *mobile lives* broadly as a mediatized and hegemonic *structure of feeling*, largely sustained by the middle classes, I theorize the inherent ambiguities of such a structure of feeling and discuss its social significance as a source of *counter-mediatization*. This leads, finally, to an assessment of how the critical bottom-up perspective, focusing on mobile lives, can help us to better understand the interplay between mediatization and globalization. I suggest that the globalizing force of mediatization is played out, and can be analysed, in relation to three registers: *connectivity*, *mobility* and *osmosity*. The triadic model offers a systematized view of how social power-geometries are (re)configured under the dual pressure of mediatization and globalization.

A critical bottom-up perspective

One of the cornerstones of this analysis has been that mediatization is an embedded as well as an embodied part of *ordinary culture*. Mediatization is not what occurs at random or beyond the regular patterns and rhythms of social life, but what occurs when people come to understand the media as *indispensable* parts of their lives. Mediatization thus evolves largely as an invisible social force. Eye-opening experiences of media indispensability tend to occur when media are actually absent (but understood as mandatory) or when access is lost; that is, when the spell of ordinariness is broken or threatened. A further cornerstone has been that the growing ordinariness of media and their expansion into new domains of social life necessarily signify a long-term *social transformation* (which does not rule out the fact that social structures and power relations are also reproduced).

In order to make these *alterations of the ordinary* accessible to empirical analysis we must identify more concrete articulations of mediatization. One way is to break mediatization down into sub-processes that can then be re-assembled into a more comprehensive view (Krotz, 2014). As Schulz (2004) argues, pre-existing forms of social practice can be *substituted* with, *accommodated* to, or *amalgamated* with processes of mediation.[1] When these sub-processes have reached a certain level of social impact they give rise to different forms of media dependence – *functional dependence, transactional dependence* and *ritual dependence* – which, in turn, correspond to the articulation of different aspects of media as cultural forms: *technics, properties* and *textures* (see Chapter 3). What we arrive at here, ultimately, is a *critical bottom-up perspective of mediatization* that can be applied across social and cultural domains (see Table 8.1).

The perspective I propose here is *bottom-up* in the sense that it aims to grasp mediatization at the level of everyday media practices and experience. Accordingly, the main arguments of this book are derived from a synthesis of fieldwork carried out in actual contexts of mediatized mobile lives. The perspective is *critical* in the sense that mediatization is conceived as a dialectical force marked by the inexorable tension between experiences of personal growth, especially related to self-realization (Honneth, 2004), and increasingly complex forms of media dependence. As the studies of mobile lives show, mediatization processes unfold in contradictory and sometimes even existentially charged ways. The appropriation of various media constitutes a pre-condition for extended mobility and spatial appropriation, but at the same time normalizes various forms of socio-material pressures, adaptations and restraints.

For example, expatriate professionals working for the United Nations (UN) in Geneva (Chapter 5) testify to hyper-mobile working conditions in which the emergence of online media platforms has made it difficult to control *which* means of communication one is expected to use, as well as the *when* and *where*. Mediatization not only reinforces their ability to stay connected while on the move, but also, through the gradual transformation of the *communicational doxa* of their field, adds elements of precariousness to their cosmopolitan identities. Similarly, the study of post-touristic urban explorers (Chapter 7) shows how the culture of connectivity as a whole (Van Dijck, 2013), and especially the expanding logic of spreadability (Jenkins et al., 2013), necessitates a growing reflexivity on behalf of individual actors in order to maintain the exclusivity of their community and to preserve the authenticity of the sites they visit. Ongoing (trans)mediatization processes have given rise to a growing sense of ambiguity, what I call *reflexive hesitation*, as well as to a polarization between different sub-fractions within the cultural field of urban exploration. Finally, the study on gentrification (Chapter 7) illustrates amongst other things how middle-class counter-urbanizers conceive of connectivity as a pre-condition

TABLE 8.1 A critical bottom-up perspective of mediatization

Sub-process of mediatization	Level of dependence	Articulated aspect of media as cultural form
Substitution	Functional dependence	Media as technics
Accommodation	Transactional dependence	Media as properties
Amalgamation	Ritual dependence	Media as textures

for their residential choices, as well as *exitability*, while at the same time constituting a threat to their sense of autonomy and *elective belonging* (Savage, 2010).

The analyses of this book also show that the mobile middle classes play a key role in problematizing the hegemonic forces that frame mediatization processes. This may seem like a puzzling conclusion, given that the middle classes play a socially normalizing role in society, legitimizing and disseminating new lifestyles, tastes and consumer demands (Chapter 4). Owing to their intermediary and relatively fluid social position, however, they also have the socio-cultural motives as well as the resources to act in transgressive, counter-hegemonic ways. The normalization of new pressures and restraints is not always and in all respects accepted but may give rise to everyday negotiations as well as outright resistance. A good example is the role of expatriate networks in gentrification processes (Chapter 7). My research indicates that these networks, sustained by connective media, often constitute a resource for middle-class expats to gain residential advantages, such as access to apartments in socially mixed inner-city neighbourhoods. At the same time these expat groups are often characterized by a cosmopolitan ethos and are eager to prob-lematize geo-social segregation as well as the insular logics of connective media. Another example concerns the differentiation of the field of urban exploration (Chapter 6). While certain fractions or groups strive to generate maximum visibility and exposure to themselves and their "brands", others try to counter the dominant imperatives of mediatization and to gain experiences marked by existential authen-ticity instead. The "handling of mediatization" is thus part and parcel of the struggle for social autonomy and recognition, and under certain conditions the countering of dominant mediatization processes works as a marker of individual autonomy and social power.

Mediatization and counter-mediatization

We should basically think about mediatization as a hegemonic force in society. The increasing need to stay connected, make oneself visible and adapt one's free time and working life to the affordances of media cannot be uncoupled from the political–economic forces of a capitalist consumer society. As stated at the outset of this book, a mediatized society is one in which the possibilities that come with new media are created not only, and probably not most significantly, as a response to pre-existing social demands, but are framed by sophisticated market logics that turn media users into producers of their own needs. For example, a new interactive game (think of Pokemon Go) may create demands for new hardware. Interactions on a social media platform may generate advertising that spurs an interest in other goods and services (think of Facebook and Google). Social media users may even start thinking of themselves as consumer brands or marketable goods that need continuous maintenance (think of Tindr).

There is nothing particularly new about these observations, of course. What they point to, above all, is that mediatization is reinforced through the synergies between market forces and our individualized culture and society (see Chapter 4). The hegemonic nature of mediatization sits well alongside other modern meta-processes, notably individualization and commercialization, which have already been given much attention within the social sciences and discussed under headings like "other-directedness" (Riesman, 1950/2001), "mobile privatization" (Williams, 1974) and "organized self-realization" (Honneth, 2004). Mediatization researchers have discussed these connections as well (see, e.g., Hjarvard, 2013: Ch. 6).

What *is* new about the current analysis is the conceptualization of mediatization as a dialectical form of social transformation in which the negative side not only produces certain forms of dependence and restraint (see Table 8.1), but also functions as a source of anti-hegemonic resistance. If we conceive of mediatization as a complex meta-process, and here I follow Krotz's (2007, 2014) work on the subject, we cannot ignore that it also *in itself* contains elements of *counter-mediatization*, which eventually may (or may not) affect the dominant directions that mediatization takes. However, we should be cautious not to interpret these dialectics in an overly mechanical way or according to any simplified thesis/anti-thesis logic. As discussed in Chapter 2, it makes better sense to conceive of media-enhanced social transformations and conflicts from a cultural materialist perspective in the spirit of Williams and Bourdieu. In this book I have used Williams's term *structure of feeling* to describe the embodied experiences that characterize mobile, and increasingly mediatized, lives. Elite cosmopolitanism, post-tourism and gentrifying lifestyles (through elective belonging) constitute concrete, intersecting manifestations of such an overarching, and basically hegemonic, structure of feeling (cf. Thrift, 1996). As we have seen in all three contexts, mobile lives harbour precisely those contradictions that Honneth (2004) associates with individual self-realization, making life an utterly ambiguous, sometimes frustrating experience. While mobile lives are closely tied to the hegemonic construction of "privilege" and popularized understandings of the "good life", they are also marked by experiences of social, cultural and existential loss. Such experiences, in turn, may foster various forms of resistance and a desire to subvert the hegemonic order in which mediatization is embedded. Expressions of counter-mediatization can thus be understood in terms of Williams's (1977) later writings on *emergent structures of feeling* (in the plural), which appear as "cultural hypotheses" (ibid.: 132) reacting to the dominant order. While such alternative movements always run the risk of being absorbed by the very same order they set out to problematize, they may also anticipate broader social and cultural transformations (see Chapter 2).

But how are we to delimit counter-mediatization? Can it refer to just any kind of "resistance"? I want to make three points regarding the general nature of counter-mediatization. First, we should conceive of counter-mediatization as *structured forms of agency*, either habitual or organized, that problematize the normalization of media dependence. Counter-mediatization should not be reduced to individual cases of media abstention or singular occasions of protest – just as mediatization refers to more than single acts of mediated communication or the mere appropriation of new media technologies – but concerns more profound transformations of people's life biographies or broader trends within a certain population or social context. If we consider the opening example of this book, in which Ruben, a UN employee in his sixties, tried to disentangle his life from media connectivities and flows by travelling less than he used to do, this tactical form of agency can possibly be seen as an embodied and deeply textured expression of counter-mediatization. It is an act whereby the individual tries to reclaim the importance of existential values and thereby regain a sense of autonomy. It is also an act whose social side effect is the risking of future opportunities to gain recognition and status within the field, thus testifying to the enduring force and dominant direction of mediatization. To the extent that this type of consequential resistance translates into a broader social formation we may even speak about a counter-hegemonic structure of feeling.

Second, counter-mediatization does not necessarily point to the rejection of media *per se*, but has to do with *the propensity among social agents to maintain a sense of autonomy* in relation to mediatization. It may involve the development of alternative ways of using media or the deliberate substitution of certain types of media for others. In my analyses of urban exploration, for example, I found that the industrial logics of connective media constituted a contested terrain that some groups developed counter-strategies to avoid, for instance through the setting up of their own, relatively independent websites. Again, such streams of counter-mediatization do not signify the end of mediatization, but should still be critically examined if we want to grasp the contradictory appearances and unexpected developments of mediatization at large. They might even be taken as signs of an emergent structure of feeling.

Third, counter-mediatization does not always emerge as a reaction to context-specific sub-processes of mediatization (even though the matrix of Table 8.1 can be helpful for analytically pinpointing what form(s) of dependence a certain instance of media resistance relates to), but may just as well evolve as *indirect responses to complex experiences of meta-processual change and/or structural conditions in society*. We have seen this in several contexts discussed in this book: middle-class counter-urbanizers whose vision is to opt out of the synergetic forces of urbanization and mediatization; female UN professionals who problematize the linkages between osmotic forms of everyday mediatization and patriarchal organizational cultures; urban explorers who oppose the commercial imperatives of urban transformations and networked media technologies.

These points highlight the fact that counter-mediatization does not refer to a homogenous type of development that in any fundamental sense escapes or converts the meta-process of mediatization. Acts of resistance are far from always successful in their ambition to achieve counter-hegemonic change. One should also avoid generalizing across populations. As discussed in Chapter 5, for example, educated middle-class women who want to advance within a transnational field (such as the UN) need to strike a balance between appropriating media in ways that enhance their flexibility in terms of work/life balance, thus increasing their chances of gaining recognition as agents within the field, and trying to subvert those (often unwritten) rules that turn media into indispensable tools for extending doxa and reproducing patriarchal structures. It takes time to change structures from within, and at the individual level extended media dependence may even be the price one has to pay to achieve social status as a woman within a traditionally male sector of society.

These are complex matters, which explains why it is difficult to discern exactly where counter-mediatization takes place, where it begins and where it ends. My point is that counter-mediatization should not only be associated with radical groups that deliberately step outside established social fields, such as activists whose very identities are built up around their resistance to dominant cultures of mediatization (including phenomena like surveillance and cultural exploitation) (see Mejias, 2013). Counter-mediatization may also arise among groups who attempt to use their social status as a resource for negotiating those forces of mediatization that foster experiences of dependence and exploitation. In such cases resistance is bound to be of a more moderate nature unless the agents are willing to risk their privileged social positions. It also means that the accumulation of valid capital is often a pre-condition for achieving the greater autonomy one needs in order to problematize mediatization. In contrast to less privileged groups the mobile middle-class fractions figuring in this book have reached positions where such opportunities gradually emerge.

In mediatized society social power is not in any clear-cut way linked to greater media access or literacy. While such resources are often the indispensable underpinnings of successful social trajectories, they are of relatively little symbolic value precisely because of their ordinariness. What is more important in the long run is the ability of individuals and groups to master the dialectic of mediatization in ways that enhance their capacity to accumulate other forms of capital and steer their life trajectories towards (organized) self-realization. The mobile lives associated with elite cosmopolitanism, post-tourism and gentrification have proven rich sources of evidence in this regard.

Mediatization, globalization and struggles for autonomy

The critical bottom-up perspective of this book has made it possible to analyse and discuss how mediatization is interwoven with other meta-processes in the cultural-material settings of everyday life. The focus on mobile middle-class lives, in turn, has been an attempt to grasp mediatization processes in contexts where people are expected to be influential agents of *individualization* and *globalization*. Elite cosmopolitanism, post-tourism and gentrification represent phenomena that are closely linked to the dominant order of organized self-realization, which in turn legitimizes (1) extended *mobility*, (2) continuous *encounters* with *Other* places and people, and (3) reflexive modes of *media connectivity* as key assets of successful life trajectories. Not surprisingly, then, agents of gentrification, elite cosmopolitanism and post-tourism are in most respects at the "forefront" of mediatization. Their lifestyles illustrate how sophisticated "media know-how" is both the outcome of and the pre-requisite for mobile lives (albeit not seen as a marker of distinction). Still, as discussed above, counter-forces to these hegemonic dynamics should not be overlooked. Growing media reliance makes the quest for fixity and disconnection just as much of a challenge as the creation of a globally oriented, or cosmopolitan, self-identity. Maintaining a sense of autonomy in relation to mediatization is thus also a way of *positioning the Self in relation to globalization.*

This statement can be more easily grasped if we break globalization down into three defining registers. First, globalization rests on the establishment of various forms of complex *connectivity* between places and people (Tomlinson, 1999). Connectivity should here be understood in a sociological sense, as links for transporting and sharing things, ideas and information, rather than being thought of in a techno-economically oriented way (cf. Van Dijck, 2013). Second, globalization is defined by the expansion and intensification of various forms of *mobility*, which are in turn dependent on the existence of infrastructural connections. As Urry (2007: Ch.8) notes, mobilities can be found on different levels, including virtual and imaginative as well as corporeal and material. Third, globalization involves the gradual absorption of foreign social and cultural elements (people, things, information and ideas) into new spaces. As suggested in Chapter 6, we can think of the intensification of such processes in terms of *osmosity*. Osmosity occurs in everyday contexts, as well as at the geo-political level where connections and flows criss-cross territorial boundaries, and typically leads to the emergence of various "inter-spaces" (ibid.: 171–80).

These three registers – connectivity, mobility and osmosity – have figured implicitly and in different guises throughout this book. The interplay between them explains how global relations, characterized by the relativized significance of location (which does not mean that *place* becomes unimportant) (Massey, 1991), expand into various domains of culture and society. As Elliot and Urry (2010) argue, privileged mobile lives can *in themselves* be seen as drivers and ideological markers of globalization. The lifestyles of "globals", by which they mean professional travellers in higher positions, are distinguished not only by intense networking and international high-speed travel in order to entertain these networks, but also "a distaste for traditional identities and communities" (ibid.: 78). Ultimately, the lives of "globals" are marked by various forms of *detached engagement*, meaning that social and professional relations are maintained largely through mediated forms of communication and a detailed *mapping of possible escape routes.*

> We are, essentially, talking about life experienced as a series of exits. From this angle, each exit is, in turn, followed by new entrances. And these entrances then entail further exits. This notion of escapism raises, of course, the thorny question of what it is, exactly, that globals are escaping from.
>
> *ibid.: 78*

This description of "globals" underlines the linkages between connectivity, mobility and osmosity. The lifestyles of "globals" contribute to the circulation of a variety of things (besides the travellers themselves) that cross borders and get absorbed into new contexts: money, goods, information, values and ideas. The characterization also helps us to see the different levels on which globalization operates. Globalization is not only a matter of material and corporeal flows, or geo-political transformations, but is also about everyday experiences, imaginations and practical adjustments – not least in relation to media and communication technologies – pertaining to individual life biographies (see also Thrift, 1996: Ch. 7; Urry, 2007: 6-7). If we look at globalization through this lens, which is more phenomenological, we discover that while the lifestyles of mobile professionals may be a driver of globalization, these "globals" can also be seen as *victims* of globalization, not in a socioeconomic sense, but from an existential and psychosocial perspective. As shown in my analysis of elite cosmopolitans (a group that largely converges with Elliot and Urry's (2010) notion of "globals"), those who aim for a transnational life biography and an affluent life on the move must also be prepared to "play the game", to obey the doxa of their fields, which means that their autonomy is bound to remain a relative matter.

It may seem odd, even provocative, to speak about *escapism* in relation to these relatively privileged professionals, especially at a time when millions of refugees are risking their lives to reach a safer place to live, but, as I argued in Chapter 5, the ongoing globalization of power geometries makes it increasingly important to problematize what we actually mean by terms like "privilege" and "eliteness" (see also Jansson, 2016). Above all, there are a number of social and existential costs associated with extensive mobility that we need to acknowledge, notably homesickness, travel fatigue and the longing for absent friends and family (Bude and Dürrschmidt, 2010). Negative side effects of mobility such as these cannot be eliminated even though you are socioeconomically affluent. Furthermore, privileged mobile lives are often marked by negative feelings linked to accentuated osmosity and connectivity, such as information overload and loss of privacy as a consequence of open-ended availability.

If we want to understand mobile lives as a structure of feeling, escapism and the rejection of a stable community base provide just one side of the picture. While the desire to entertain a cosmopolitan self-identity, ultimately feeling at home in the world, or in the movement as such, is an important driver behind many mobile life biographies, there are also factors that pull individuals back, making them doubt the deeper meaning of their trajectories and the sacrifices they make. Many mobile professionals, for example, struggle with ambiguous experiences of stress and frustration when they realize the incompatibility of different life goals. As Fast and Lindell (2016) show, the typical business traveller is often much more home-oriented than one might imagine, seeing mobility as a necessary evil for maintaining a professional career. Similarly, my analysis of elective belonging points to the common intertwining of mobile, or global, life aspirations with a desire to establish deep emotional attachments to a particular place, what I call *creative topophilia* (Chapter 6). We should thus keep in mind that globalization also produces large groups of gentrifiers, counter-urbanizers, home-comers and other cosmopolitan cohorts, who settle in one locality, or return to their places of origin, or in other ways swap their mobile lives for more place-based lifestyles (Dürrschmidt, 2016).

What is the role of media in all this? While the analyses of this book cover a relatively diverse sample of mobile lives (and not just "globals"), they still offer, I believe, a coherent view of how the appropriation of media sustains further globalization. For example, urban explorers entertain their networks (globally as well as locally) and plan their trips by using various online communities and social media platforms through which they also circulate representations of sites they have visited, and which may then spur other explorers to follow. Urban explorers also generate and circulate aestheticized views of previously unseen urban landscapes that have a globalizing effect within wider realms of society as they are taken up by sectors like tourism and popular culture. Similarly, the media practices of gentrifying groups and middle-class expats are marked by far-ranging connectivities that provide *access as well as exitability* vis-à-vis certain places, and open up work places and home-places to a variety of foreign impulses, ultimately problematizing the boundaries between home and away, work and leisure, fixity and flow. All this can be understood as the *mediated intensification of connectivities, mobilities and osmosities.*

This intensification makes everyday lifeworlds not just globalized, but also *mediatized.* New media dependences emerge in relation to each of the three registers of globalization, most often resulting in mutual reinforcement. However, as my studies have also shown, obedience to the globalizing force of mediatization can never be fully married to the ideal of self-realization. In mediatized social spaces the ability to master complex media resources *per se* is rarely recognized as a sign of social status. Rather, it is taken for granted as a pre-condition for reaching other goals, and is thus not expected to take centre-stage or dominate over other assets, or forms of capital, that are deemed valid within a given setting. As we saw in the study of UN expats, connective media practices do not *replace* mobility and face-to-face meetings. Similarly, in the context of urban exploration exaggerated media use was associated with the erosion of artistic as well as existential forms of authenticity. Among provincial gentrifiers connectivity was typically seen as a pre-condition for settlement even though their prime residential goal in many cases was to get away from information flows and regain a sense of control over connectivity and osmosity.

Such examples of socio-cultural devaluation bring us back to the question of counter-mediatization and social power. They underscore that social power rests on the agent's ability to play the game and to calculate how best to use various resources in ways that express belonging to the field or group while at the same time marking out a significant degree of autonomy. As Kaufmann (2002) suggests with his concept of *motility*, referring to *the ability to control one's own mobility*, the capacity to decide when to be mobile and when to stay put is a stronger marker of social power than mobility as such. The same thing, I argue, goes for connectivity and osmosity. Social power is associated with the agent's ability to master and prioritize which connections should be sustained and to manage the filters of the lifeworld in order not to become enslaved by various information flows and material pressures.

★ ★ ★

These insights regarding mobile lives, I argue, provide a good starting point, a kind of epicentre, for exploring a whole set of other trends that influence and demarcate contemporary culture and society. Among such trends we may identify phenomena like life-coaching, media abstention communities and various therapeutic discourses of self-help and self-surveillance. We may also think of new online services and media applications designed for the purpose of making people better equipped to manage their hurried lives and work out their life puzzles. What these phenomena have in common, in spite of their contradictory appearance, is that they are legitimized in relation to a broadly shared middle-class experience – a structure of feeling – of mediatized and mobile lives. While they may look like instances of counter-mediatization (and in some cases there is an affinity with more radical movements), they should be seen above all as evidence of how hegemonic orders absorb social ambiguities and turn them into the raw material for new services and products. In other words, they are examples of ideological forms of recognition (Honneth, 2012). This is not the place to pursue any further analyses of these phenomena. Because of their social, cultural and economic pervasiveness, however, they mark out important areas to study if we want to deepen our understanding of the social consequences of mediatization in contemporary capitalist societies.

Note

1   Schulz (2004) also includes "extension" as a sub-process of mediatization, but I suggest we should see that rather as an instance of mediation (see also Jansson, 2015b).

REFERENCES

Abe, K. (2009). The myth of media interactivity: Technology, communications and surveillance in Japan, *Theory, Culture and Society* 26 (2–3): 73–88.

Abrams, J., Bliss, J. and Gosnell, H. (2013). Reflexive gentrification of working lands in the American West: Contesting the "middle landscape", *Journal of Rural and Community Development* 8(3): 144–58.

Acland, C.R. (ed.) (2007). *Residual Media*. Minneapolis, MN: University of Minnesota Press.

Adams, P.C. (1992). Television as gathering place, *Annals of the Association of American Geographers* 82(1): 117–35.

Adams, P.C., Hoelscher, S. and Till, K.E. (2001). Place in context: Rethinking humanist geographies. In Adams, P.C., Hoelscher, S. and Till, K.E. (eds.), *Textures of Place: Exploring Humanist Geographies*. Minneapolis: University of Minnesota Press.

Adolf, M. (2014). Involuntaristische Mediatisierung. Big Data als Herausforderung einer informationalisierten Gesellschaft. In Ortner, H., Pfurtscheller, D., Rizzolli, M. and Wiesinger, A. (eds.) *Datenflut und Informationskanäle*. Innsbruck: Innsbruck University Press.

Albrechtslund, A.-M. and Albrechtslund, A. (2014). Social media as leisure culture, *First Monday* 19(4).

Allmer, T. (2015). *Critical Theory and Social Media: Between Emancipation and Commodification*. London: Routledge.

Allmer, T., Fuchs, C., Kreilinger, V. and Sevignani, S. (2014). Social networking sites in the surveillance society: Critical perspectives and empirical findings. In Jansson, A. and Christensen, M. (eds.) *Media, Surveillance and Identity: Social Perspectives*. New York: Peter Lang.

Allon, F. (2004). An ontology of everyday control: Space, media flows and "smart" living in the absolute present. In Couldry, N. and McCarthy, A. (eds.) *Mediaspace: Place, Scale and Culture in a Media Age*. London: Routledge.

Amin, A. (2002). Ethnicity and the multicultural city: Living with diversity, *Environment and Planning A* 34(6): 959–80.

Amin, A. (2012). *Land of Strangers*. Cambridge: Polity Press.

Amin, A. and Thrift, N. (2007). Cultural-economy and cities, *Progress in Human Geography* 31(2): 143–61.

Andersson, M. (2013). Multi-contextual lives: Transnational identifications under mediatized conditions, *European Journal of Cultural Studies* 16(4): 387–404.

Andrejevic, M. (2007). *iSpy: Surveillance and Power in the Interactive Era*. Lawrence: University Press of Kansas.

Andrejevic, M. (2013). *Infoglut: How Too Much Information Is Changing the Way We Think and Know*. London: Routledge.

Andrejevic, M. (2014). The infinite debt of surveillance in the digital economy. In Jansson, A. and Christensen, M. (eds.) *Media, Surveillance and Identity: Social Perspectives.* New York: Peter Lang.

Appadurai, A. (1990). Disjuncture and difference in the global cultural economy, *Theory, Culture and Society* 7(2–3): 295–310.

Asp, K. (2014). Mediatization: Rethinking the question of media power. In Lundby, K. (ed.) *Mediatization of Communication (Handbooks of Communication Science, Vol. 22)*. Berlin: De Gruyter Mouton.

Bakardjieva, M. and Gaden, G. (2012). Web 2.0 technologies of the self, *Philosophy & Technology* 25(3): 399–413.

Bankovsky, M. and Le Goff, A. (eds.) (2012). *Recognition Theory and Contemporary French Moral and Political Philosophy: Reopening the Dialogue*. Manchester: Manchester University Press.

Barnett, C. (2005). Ways of relating: Hospitality and the acknowledgement of otherness, *Progress in Human Geography* 29(1): 5–21.

Baudrillard, J. (1972/1981). *For a Critique of the Political Economy of the Sign*. St Louis, MO: Telos.

Baym, N.K. (2010). *Personal Connections in the Digital Age*. Cambridge: Polity Press.

Baym, N.K. and Boyd, D. (2012). Socially mediated publicness: An introduction, *Journal of Broadcasting and Electronic Media* 56(3): 320–29.

Beaverstock, J.V. (2002). Transnational elites in global cities: British expatriates in Singapore's financial district, *Geoforum* 33: 525–38.

Beck, U. (2004/2006). *The Cosmopolitan Vision*. Cambridge: Polity Press.

Beck, U. and Beck-Gernsheim, E. (2002). *Individualization: Institutionalized Individualism and Its Social and Political Consequences.* London: Sage.

Bengtsson, S. (2011). Imagined user modes: Media morality in everyday life, *International Journal of Cultural Studies* 15(2): 181–96.

Bengtsson, S. (2015a). An ethics of ambiguity in a culture of connectivity? Paper presented at the international research workshop "Mediatisation of Culture and Everyday Life", 23–24 April, Stockholm, Sweden.

Bengtsson, S. (2015b). Digital distinctions: Mechanisms of difference in digital media use, *MedieKultur: Journal of Media and Communication Research* 31(58): 30–48.

Bennett, L. (2011). Bunkerology – a case study in the theory and practice of urban exploration, *Environment and Planning D: Society and Space* 29(3): 421–39.

Benson, M. (2014). Trajectories of middle-class belonging: The dynamics of place attachment and classed identities, *Urban Studies* 51(14): 3097–112.

Benson, M. (2016). Deconstructing belonging in lifestyle migration: Tracking the emotional negotiations of the British in rural France, *European Journal of Cultural Studies* 19(5): 481–94.

Benson, M. and O'Reilly, K. (2009). Migration and the search for a better way of life: A critical exploration of lifestyle migration, *The Sociological Review* 57(4): 608–25.

Berger, P., Berger, B. and Kellner, H. (1973). *The Homeless Mind. Modernization and Consciousness*. New York: Vintage Books.

Blau, P.M. (1994). *Structural Contexts of Opportunities*. Chicago: University of Chicago Press.

Bolin, G. (2014). Institution, technology, world: Relationships between the media, culture, and society. In Lundby, K. (ed.) *Mediatization of Communication (Handbooks of Communication Science, Vol. 22)*. Berlin: De Gruyter Mouton.

Bolin, G. (2017). *Media Generations: Experience, Identity and Mediatised Social Change.* London: Routledge.

Boltanski, L. (1996/1999). *Distant Suffering: Morality, Media and Politics.* Cambridge: Cambridge University Press.

Boltanski, L. and Chiapello, E. (1999/2007). *The New Spirit of Capitalism.* London: Verso.

Bourdieu, P. (1965/1990). *Photography: A Middle-Brow Art.* Cambridge: Polity Press.

Bourdieu, P. (1972/1977). *Outline of a Theory of Practice.* Cambridge: Cambridge University Press.

Bourdieu, P. (1979/1984). *Distinction: A Social Critique of the Judgment of Taste.* London: Routledge.

Bourdieu, P. (1980/1990). *The Logic of Practice.* Cambridge: Polity Press.

Bourdieu, P. (1983). The field of cultural production, or: The economic world reversed, *Poetics* 12: 311–56.

Bourdieu, P. (1996). *The State Nobility: Elite Schools in the Field of Power.* Cambridge: Polity Press.

Bourdieu, P. (1997/2000). *Pascalian Meditations.* London: Polity Press.

Bourdieu, P. and Wacquant, L.J.D. (1992). *An Invitation to Reflexive Sociology.* Chicago: University of Chicago Press.

Boyd, D. and Marwick, A. (2011). I tweet honestly, I tweet passionately: Twitter users, context collapse, and the imagined audience. *New Media & Society* 13(1): 114–133.

Broadbent, S. (2011). *L'Intimité au Travail.* Paris: FYP Editions.

Bude, H. and Dürrschmidt, J. (2010). What's wrong with globalization? Contra flow-speak: Towards an existential turn in globalization theory, *European Journal of Social Theory* 13(4): 481–500.

Bühlmann, F., David, T. and Mach, A. (2013). Cosmopolitan capital and the internationalization of the field of business elites: Evidence from the Swiss case, *Cultural Sociology* 7(2): 211–29.

Bull, M. (2001). The world according to sound: Investigating the world of Walkman users, *New Media and Society* 3(2): 179–97.

Bull, M. (2007). *Sound Moves: iPod Culture and Urban Experience.* London: Routledge.

Butler, T. (2003). Living in the bubble: Gentrification and its "others" in London, *Urban Studies* 40(12): 2469–86.

Butler, T. and Lees, L. (2006). Super-gentrification in Barnsbury, London: Globalization and gentrifying global elites at the neighbourhood level, *Transactions of the Institute of British Geographers* 31(4): 467–87.

Calhoun, C. (2002). The class consciousness of frequent travelers: Toward a critique of actually existing cosmopolitanism, *The South Atlantic Quarterly* 101(4): 869–97.

Carroll, W.K. (2009). Transnationalists and national networkers in the global corporate elite, *Global Networks* 9(3): 289–314.

Castells, M. (1996). *The Rise of the Network Society.* Oxford: Blackwell.

Caulfield, J. (1994). *City Form and Everyday Life: Toronto's Gentrification and Critical Social Practice.* Toronto: University of Toronto Press.

Chadwick, A. (2013). *The Hybrid Media System.* Oxford: Oxford University Press.

Christensen, M. and Jansson, A. (2012). Fields, territories and bridges: Networked communities and mediated surveillance in transnational social space. In Fuchs, C., Boersma, K., Albrechtslund, A. and Sandoval, M. (eds.) *The Internet and Surveillance: The Challenges of Web 2.0 and Social Media.* London: Routledge.

Christensen, M. and Jansson, A. (2015). *Cosmopolitanism and the Media: Cartographies of Change*. Basingstoke: Palgrave Macmillan.

Christensen, T.H. (2009). "Connected presence" in distributed family life, *New Media and Society* 11(3): 433–51.

Clark, L.S. (2009). Mediatization: Where media ecology meets cultural studies. In Lundby, K. (ed.) *Emerging Theories of Mediatization*. London: Routledge.

Couldry, N. (2003a). *Media Rituals: A Critical Approach*. London: Routledge.

Couldry, N. (2003b). Media meta-capital: Extending the range of Bourdieu's field theory. *Theory and Society* 32 (5–6): 653–77.

Couldry, N. (2012). *Media, Society, World: Social Theory and Digital Media Practices*. Cambridge: Polity Press.

Couldry, N. (2014). Mediatization and the future of field theory. In Lundby, K. (ed.) *Mediatization of Communication (Handbooks of Communication Science, Vol. 21)*. Berlin: De Gruyter Mouton.

Couldry, N. and Hepp, A. (2013). Conceptualizing mediatization: Contexts, traditions, arguments, *Communication Theory* 13(3): 191–202.

Couldry, N. and McCarthy, A. (eds.) (2004). *Media Space: Place, Scale and Culture in a Media Age*. London: Routledge.

Crampton, J. (1995). The ethics of GIS, *Cartography and Geographic Information Systems* 22(1): 84–89.

Cresswell, T. (2006). *On the Move: Mobility in the Modern Western World*. London: Routledge.

Daenekindt, S. and Roose, H. (2014). Social mobility and cultural dissonance, *Poetics* 42: 82–97.

Danielsson, M. (2014). Digitala distinktioner: Klass och kontinuitet i unga mäns vardagliga mediepraktiker. PhD dissertation, Jönköping University.

Davidson, M. and Lees, L. (2005). New-build "gentrification" and London's riverside renaissance, *Environment and Planning A* 37(7): 1165–90.

De Silvey, C. and Edensor, T. (2012). Reckoning with ruins, *Progress in Human Geography* 37(4): 465–85.

Deacon, D. and Stanyer, J. (2014). Mediatization: Key concept or conceptual bandwagon? *Media, Culture & Society* 36(7): 1032–44.

Deacon, D. and Stanyer, J. (2015). "Mediatization *and*" or "mediatization *of*"? A response to Hepp et al., *Media, Culture & Society* 37(4): 655–57.

Dean, J. (2009). *Democracy and Other Neoliberal Fantasies: Communicative Capitalism and Left Politics*. Durham, NC: Duke University Press.

Delanty, G. (2009). *The Cosmopolitan Imagination: The Renewal of Critical Social Theory*. Cambridge: Cambridge University Press.

Derrida, J. (2000). Hospitality, *Angelaki* 5: 3–18.

Derrida, J. (2001). *On Cosmopolitanism and Forgiveness*. London: Routledge.

Derrida, J. (2005). The principle of hospitality, *Parallax* 11(1): 6–9.

Deuze, M. (2011). Media life, *Media, Culture & Society* 33(1): 137–48.

Deuze, M. (2012). *Media Life*. Cambridge: Polity Press.

Dikeç, M. (2002). Pera, peras, poros: Longing for spaces of hospitality, *Theory, Culture and Society* 19: 227–47.

Dikeç, M. (2009). Justice and the spatial imagination. In Marcuse, P., Connolly, J., Novy, J., Olivo, I., Potter C. and Steil, J. (eds.) *Searching for the Just City: Debates in Urban Theory and Practice*. London: Routledge.

Dodge, M. and Kitchin, R. (2006). Exposing the secret city: urban exploration as "space hacking". Paper presented at the AAG Annual Meeting, Chicago, March.

Doucet, B. (2014). A process of change and a changing process: Introduction to the special issue on contemporary gentrification, *Tijdschrift voor Economische en Sociale Geografie* 105(2): 125–39.

Dürrschmidt, J. (2016). The irresolvable unease about be-longing: Exploring globalized dynamics of homecoming, *European Journal of Cultural Studies* 19(5): 495–510.

Edensor, T. (2005). The ghosts of industrial ruins: Ordering and disordering memory in excessive space, *Environment and Planning D: Society and Space* 23: 829–49.

Edensor, T. (2007). Sensing the ruin, *Senses & Society* 2(2): 217–32.

Ekström, M., Fornäs, J., Jansson, A. and Jerslev, A. (2016). Three tasks for mediatization research: Contributions to an open agenda, *Media, Culture & Society* 38(7): 1090–1108.

Elliot, A. (2014). Elsewhere: Tracking the mobile lives of globals. In Birtchnell, T. and Caletrío, J. (eds.) *Elite Mobilities*. London: Routledge.

Elliot, A. and Urry, J. (2010). *Mobile Lives*. London: Routledge.

Elwood, S. and Leszczynski, A. (2012). New spatial media, new knowledge politics, *Transactions of the Institute of British Geographers* 38: 544–59.

Emmison, M. (2003). Social class and cultural mobility: reconfiguring the cultural omnivore thesis, *Journal of Sociology* 39(3): 211–30.

Eriksson Baaz, M. (2005). *The Paternalism of Partnership: A Postcolonial Reading of Identity in Development Aid*. London: Zed Books.

Esser, F. and Strömbäck, J. (eds.) (2014). *Mediatization of Politics: Understanding the Transformation of Western Democracies*. Basingstoke: Palgrave Macmillan.

Fast, K. (2015). World-building vs. brand-building: Transformers as a Marvel outcast and Hollywood star. Paper presented at the 65th ICA Annual Conference, San Juan, Puerto Rico, 21–25 May.

Fast, K. and Lindell, J. (2016). The elastic mobility of business elites – negotiating the "Home" and "Away" continuum, *European Journal of Cultural Studies* 19(5): 435–49.

Featherstone, M. (1991). *Consumer Culture and Postmodernism*. London: Sage.

Feifer, M. (1985). *Going Places*. London: Macmillan.

Felgenhauer, T. and Quade, D. (2012). Society and geomedia – some reflections from a social theory perspective. In Jekel, T., Car, A., Strobl, J. and Griesebner, G. (eds.) *GI_Forum 2012: Geovisualization, Society and Learning*. Berlin: VDE Verlag.

Findahl, O. (2014). *Svenskarna och Internet 2014*. Stockholm: SE (Stiftelsen för Internetinfrastruktur).

Florida, R. (2002). *The Rise of the Creative Class*. New York: Basic Books.

Fornäs, J. (2013). The dialectics of communicative and immanent critique in cultural studies, *tripleC* 11(2): 504–14.

Fraser, N. (1997). *Justice Interruptus: Critical Reflections on the "Postsocialist" Condition*. New York: Routledge.

Fraser, N. (2000a). Why overcoming prejudice is not enough: A rejoinder to Richard Rorty, *Critical Horizons* 1(1): 21–28.

Fraser, N. (2000b). Rethinking recognition, *New Left Review* 3 (May–June): 107–20.

Fraser, N. (2001). Recognition without ethics?, *Theory, Culture and Society* 18(2–3): 21–42.

Fuchs, C. (2014). *Social Media: A Critical Introduction*. London: Sage.

Gansing, K. (2013). Transversal Media Practices: Media Archaeology, Art and Technological Development. PhD dissertation, Malmö University.

Garcia, L.-M. (2016). Techno-tourism and post-industrial neo-romanticism in Berlin's electronic dance music scenes, *Tourist Studies* 16(3): 276–95.

Garrett, B.L. (2010). Urban explorers: quests for myth, mystery and meaning, *Geography Compass* 4(10): 1448–61.

Garrett, B.L. (2012). Place Hacking: Tales of Urban Exploration. PhD dissertation, London, Royal Holloway.

Garrett, B.L. (2014a). Undertaking recreational trespass: urban exploration and infiltration, *Transactions of the Institute of British Geographers* 39(1): 1–13.

Garrett, B.L. (2014b). *Explore Everything: Place-Hacking the City*. London: Verso.

Georgiou, M. (2013). *Media and the City: Cosmopolitanism and Difference*. Cambridge: Polity Press.

Giddens, A. (1984). *The Constitution of Society*. Cambridge: Polity.

Giddens, A. (1991). *Modernity and Self-Identity: Self and Society in the Late Modern Age*. Cambridge: Polity Press.

Gillespie, T. (2010). The politics of "platforms", *New Media & Society* 12(3): 347–64.

Gin, J. and Taylor, D.E. (2010). Movements, neighborhood change, and the media: Newspaper coverage of anti-gentrification activity in the San Francisco Bay Area: 1995–2005, *Research in Social Problems and Public Policy* 18: 1–26.

Gitelman, L. and Pingree. G. (2003). What is new about new media? In Gitelman, L. and Pingree, G. (eds.) *New Media 1740–1915*. Cambridge, MA: MIT Press.

Glass, R. (1964). Introduction: Aspects of change. In Glass, R. (ed.) *London: Aspects of Change*. London: MacGibbon and Kee.

Glick-Schiller, N., Darieva, T. and Gruner-Domic, S. (2011). Defining cosmopolitan sociability in a transnational age: An introduction, *Ethnic and Racial Studies* 34(3): 399–418.

Glick-Schiller, N. and Salazar, N. (2013). Regimes of mobility across the globe, *Journal of Ethnic and Migration Studies* 39(2): 183–200.

Goffman, E. (1959). *The Presentation of Self in Everyday Life*. Harmondsworth: Penguin.

Gordon, A. (1997). *Ghostly Matters*. Minneapolis, MN: University of Minnesota Press.

Graham, S. (2004). The software-sorted city: Rethinking the "digital divide". In Graham, S. (ed.) *The Cybercities Reader*. London: Routledge.

Graham, S. (2005). Software-sorted geographies, *Progress in Human Geography* 29 (5): 562–80.

Graham, S. and Marvin, S. (2001). *Splintering Urbanism: Networked Infrastructures, Technological Mobilities and the Urban Condition*. London: Routledge.

Gregg, M. (2008). The normalisation of flexible female labour in the information economy. *Feminist Media Studies* 8(3): 285–99.

Gregg, M. (2011). *Work's Intimacy*. Cambridge: Polity Press.

Grusin, R. (2010). *Premediation: Affect and Mediality after 9/11*. Basingstoke: Palgrave Macmillan.

Gustafsson, P. (2009). More cosmopolitan, no less local: The orientations of international travellers, *European Societies* 11(1): 25–47.

Halfacree, K. (2007). Trial by space for a "radical turn": Introducing alternative localities, representations and lives, *Journal of Rural Studies* 23: 125–41.

Halfacree, K. (2010). Reading rural consumption practices for difference: Bolt-holes, castles and life-rafts, *Culture Unbound* 2: 241–63.

Hall, J.A. and Baym, N.K. (2012). Calling and texting (too much): Mobile maintenance expectations, (over)dependence, entrapment, and friendship satisfaction, *New Media and Society* 14(2): 316–31.

Hannerz, U. (1990). Cosmopolitans and locals in a world culture, *Theory, Culture and Society* 7(2): 237–51.

Hannerz, U. (1992). *Cultural Complexity: Studies in the Social Organization of Meaning*. New York: Columbia University Press.

Hartmann, M. (2006). The triple articulation of ICTs: Media as technological objects, symbolic environments and individual texts. In Berker, T., Hartmann, M., Punie, Y. and Ward, K.J. (eds.) *The Domestication of Media and Technology*. Maidenhead, Berkshire: Open University Press.

Harvey, D. (2012). *Rebel Cities: From the Right to the City to the Urban Revolution*. London: Verso.

Hegel, G. (1977). *Hegel's Phenomenology of Spirit*. Oxford: Oxford University Press.

Hepp, A. (2009). Localities of diasporic communicative spaces: Material aspects of translocal mediated networking, *Communication Review* 12(4): 327–48.

Hepp, A. (2013). *Cultures of Mediatization*. Cambridge: Polity Press.

Hepp, A., Hjarvard, S. and Lundby, K. (2015). Mediatization: theorizing the interplay between media, culture and society, *Media, Culture & Society* 37(2): 314–24.

High, S. and Lewis, D.W. (2007). *Corporate Wasteland: The Landscape and Memory of Deindustrialization*. Ithaca, NY: Cornell University Press.

Hillis, K. (2009). *On Line a Lot of the Time: Ritual, Fetish, Sign*. Durham, NC: Duke University Press.

Hines, J.D. (2012). The post-industrial regime of production/consumption and the rural gentrification of the New West archipelago, *Antipode* 44(1): 74–97.

Hirsch, E. (1992). The long term and the short term of domestic consumption: An ethnographic case study. In Silverstone, R. and Hirsch, E. (eds.) *Consuming Technologies: Media and Information in Domestic Spaces*. London: Routledge.

Hjarvard, S. (2008). The mediatization of society: A theory of the media as agents of social and cultural change, *Nordicom Review* 29(2): 105–34.

Hjarvard, S. (2013). *The Mediatization of Culture and Society*. London: Routledge.

Hjarvard, S. (2014). Mediatization and cultural and social change: An institutional perspective. In Lundby, K. (ed.) *Mediatization of Communication (Handbooks of Communication Science, Vol. 21)*. Berlin: De Gruyter Mouton.

Honneth, A. (2004). Organized self-realization: Some paradoxes of individualization, *European Journal of Social Theory* 7(4): 463–78.

Honneth, A. (2009). *Pathologies of Reason: On the Legacy of Critical Theory*. New York: Columbia University Press.

Honneth, A. (2012). *The I in the We: Studies in the Theory of Recognition*. London: Polity Press.

Hudson, J. (2014). The affordances and potentialities of derelict urban spaces. In Olsen, B. and Petursdottir, T. (eds.) *Ruin Memories: Materiality, Aesthetics and the Archaelogy of the Recent Past*. London: Routledge.

Hutchins, B. (2016). "We don't need no stinking smartphones!" Live stadium sports events, mediatization, and the non-use of mobile media, *Media, Culture & Society* 38(3): 420–36.

Igarashi H. and Saito, H. (2014). Cosmopolitanism as cultural capital: Exploring the intersection of globalization, education and stratification, *Cultural Sociology* 8(3): 222–39.

Ihde, D. (1990). *Technology and the Lifeworld: From Garden to Earth*. Bloomington, IN: Indiana University Press.

Ihde, D. (1999). *Expanding Hermeneutics: Visualism in Science. Evanston*, Evanston, IL: Northwestern University Press.

Ihlen, Ø. and Pallas, J. (2014). Mediatization of corporations. In Lundby, K. (ed.) *Mediatization of Communication (Handbooks of Communication Science, Vol. 21)*. Berlin: De Gruyter Mouton.

Jansson, A. (2001). Image Culture: Media, Consumption and Everyday Life in Reflexive Modernity. PhD dissertation, Department of Journalism and Mass Communication, Gothenburg University.

Jansson, A. (2002a). Spatial phantasmagoria: The mediatization of tourism experience, *European Journal of Communication* 17(4): 429–43.

Jansson, A. (2002b). The mediatization of consumption: Towards an analytical framework of image culture, *Journal of Consumer Culture* 2(1): 5–31.

Jansson, A. (2003a). Lifeworlds under siege? A study of mediatization as intrusion and restraint, *Nordicom Review* 24(1): 3–17.

Jansson, A. (2003b). The negotiated city image: Symbolic reproduction and change through urban consumption, *Urban Studies* 39(3): 463–79.

Jansson, A. (2005). Re-encoding the spectacle: Urban fatefulness and mediated stigmatization in "The City of Tomorrow", *Urban Studies* 41(10): 1671–91.

Jansson, A. (2006). Textural analysis: Materialising media space. In Falkheimer, J. and Jansson, A. (eds.) *Geographies of Communication: The Spatial Turn in Media Studies.* Gothenburg: Nordicom.

Jansson, A. (2007a). A sense of tourism: New media and the dialectic of encapsulation/decapsulation, *Tourist Studies* 7(1): 5–24.

Jansson, A. (2007b). Encapsulations: The production of a future gaze at Montreal's Expo 67, *Space and Culture* 10(4): 418–36.

Jansson, A. (2007c). Texture: A key concept for communication geography, *European Journal of Cultural Studies* 10(2): 185–202.

Jansson, A. (2009). Mobile belongings: Texture and stratification in mediatization processes. In Lundby, K. (ed.) *Mediatization: Concept, Changes, Consequences.* New York: Peter Lang.

Jansson, A. (2010a). The city in-between: Communication geographies, tourism and the urban unconscious. In Knudsen, B.T. and Waade, A.-M. (eds.) *Re-Investing Authenticity: Tourism, Place and Emotions.* Bristol: Channel View Publications.

Jansson, A. (2010b). Mediatization, spatial coherence and social sustainability: The role of digital media networks in a Swedish countryside community, *Culture Unbound* 2: 177–92.

Jansson, A. (2011). Cosmopolitan capsules: Mediated networking and social control in expatriate spaces. In Christensen, M., Jansson, A. and Christensen, C. (eds.) *Online Territories: Globalization, Mediated Practice and Social Space.* New York: Peter Lang.

Jansson, A. (2012). Perceptions of surveillance: Reflexivity and trust in a mediatized world (the case of Sweden), *European Journal of Communication* 27(4): 410–27.

Jansson, A. (2013a). Mediatization and social space: Reconstructing mediatization for the transmedia age, *Communication Theory* 23(3): 279–96.

Jansson, A. (2013b). The hegemony of the urban/rural divide: Cultural transformations and mediatized moral geographies in Sweden, *Space and Culture* 15(1): 88–103.

Jansson, A. (2014). Textures of interveillance: A socio-material approach to the integration of transmedia technologies in domestic life. In Jansson, A. and Christensen, M. (eds.) *Media, Surveillance and Identity: Social Perspectives.* New York: Peter Lang.

Jansson, A. (2015a). The molding of mediatization: The stratified indispensability of media in close relationships, *Communications* 40(4): 379–401.

Jansson, A. (2015b). Interveillance: A new culture of recognition and mediatization, *Media and Communication* 3(3): 81–90.

Jansson, A. (2016). Mobile elites: Understanding the ambiguous lifeworlds of sojourners, dwellers and homecomers, *European Journal of Cultural Studies* 19(5): 421–34.

Jansson, A. and Andersson, M. (2012). Mediatization at the margins: Cosmopolitanism, network capital and spatial transformation in rural Sweden, *Communications* 37(2): 173–94.

Jansson, A. and Lindell, J. (2015). News media consumption in the transmedia age: Amalgamations, orientations and geo-social structuration, *Journalism Studies* 16(1): 79–96.

Jenkins, H. (2006). *Convergence Culture: Where Old and New Media Collide.* New York: New York University Press.

Jenkins. H., Ford, S. and Green, J. (2013). *Spreadable Media: Creating Value and Meaning in a Networked Culture.* New York: New York University Press.

Johnson, P.C. (2014). Cultural literacy, cosmopolitanism and tourism research, *Annals of Tourism Research* 44: 255–69.

Jörnmark, J. (2007). *Övergivna Platser*. Lund: Historiska Media.

Jörnmark, J. (2008). *Övergivna Platser Två*. Lund: Historiska Media.

Kant, I. (1795/2003). *On Perpetual Peace: A Philosophical Sketch*. Indianapolis, IN: Hackett Publishing.

Kantola, A. (2014). Mediatization of power: Corporate CEOs in flexible capitalism, *Nordicom Review* 35(2): 29–41.

Kaufmann, V. (2002). *Re-thinking Mobility: Contemporary Sociology*. Aldershot: Ashgate.

Kaun, A. (2014). I really don t like them! Exploring citizens' media criticism, *European Journal of Cultural Studies* 17(5): 489–506.

Kaun, A. and Schwarzenegger, C. (2014). No media, less life? Online disconnection in mediatized worlds, *First Monday* 19(11).

Kennedy, P. (2009). The middle-class cosmopolitan journey: The life trajectories and transnational affiliations of skilled EU migrants in Manchester. In Nowicka, M. and Rovisco, M. (eds.) *Cosmopolitanism in Practice*. Farnham: Ashgate.

Kern, S. (2003). *The Culture of Time and Space, 1880–1918*. 2nd edition. Cambridge, MA: Harvard University Press.

Kitchin, R. and Dodge, M. (2011). *Code/Space: Software and Everyday Life*. Cambridge, MA: MIT Press.

Klausen, M. (2012). Making place in the media city, *Culture Unbound* 4: 559–77.

Klausen, M. (forthcoming). The urban exploration imagery: Mediatization, commodification and affect, *Space and Culture*.

Krajina, Z., Moores, S. and Morley, D. (2014). Non-media-centric media studies: A cross-generational conversation, *European Journal of Cultural Studies* 17(6): 682–700.

Krotz, F. (2007). The meta-process of "mediatization" as a conceptual frame, *Global Media and Communication* 3(3): 256–60.

Krotz, F. (2014). Mediatization as a mover in modernity: social and cultural change in the context of media change. In Lundby, K. (ed.) *Mediatization of Communication (Handbooks of Communication Science, Vol. 22)*. Berlin: De Gruyter Mouton.

Lahire, B. (2008). The individual and the mixing of genres: cultural dissonance and self-distinction, *Poetics* 36(2): 166–88.

Lamont, M. and Aksartova, S. (2002). Ordinary cosmopolitanisms: Strategies for bridging racial differences among working class men, *Theory, Culture & Society* 19(4): 1–25.

Lapenta, F. (2011). Geomedia: On location-based media, the changing status of collective image production and the emergence of social navigation systems, *Visual Studies* 26(1): 14–24.

Lash, S. and Urry, J. (1994). *Economies of Signs and Space*. London: Sage.

Lawler, S. (2005). Disgusted subjects: the making of middle-class identities, *The Sociological Review* 53(3): 429–46.

Lees, L. (2000). A reappraisal of gentrification: Towards a "geography of gentrification", *Progress in Human Geography* 24(3): 389–408.

Lees, L. (2003). Super-gentrification: The case of Brooklyn Heights, New York City, *Urban Studies* 40(12): 2487–2509.

Lefebvre, H. (1968/1993). The right to the city. In Ockman, J. (ed.) *Architecture Culture: A Documentary Anthology*. New York: Rizzoli International Publications.

Lefebvre, H. (1971/1984). *Everyday Life in the Modern World*. London: Continuum.

Lefebvre, H. (1974/1991). *The Production of Space*. Oxford: Blackwell.

Levi-Strauss, C. (1963). *Structural Anthropology*. New York: Basic Books.

Ley, D. (1996). *The New Middle Class and the Remaking of the Central City*. Oxford: Oxford University Press.

Ley, D. (2003). Artists, aestheticisation and the field of gentrification, *Urban Studies* 40(12): 2527–44.

Licoppe, C. (2004). "Connected" presence: The emergence of a new repertoire for managing social relationships in a changing communication technoscape, *Environment and Planning D: Society and Space* 22: 125–56.

Lindell, J. (2014). *Cosmopolitanism in a Mediatized World: The Social Stratification of Global Orientations*. Karlstad: Faculty of Arts and Sciences, Karlstad University.

Ling, R. (2008). *New Tech, New Ties: How Mobile Communication Is Reshaping Social Cohesion*. Cambridge, MA: MIT Press.

Livingstone, S. (2007). On the material and the symbolic: Silverstone's double articulation of research traditions in new media studies, *New Media and Society* 9(1): 16–24.

Lizardo, O. (2006). How cultural tastes shape personal networks, *American Sociological Review* 71(5): 778–807.

Löfgren, O. (2009). Domesticated media: Hiding, dying or haunting. In Jansson, A. and Lagerkvist, A. (eds.) *Strange Spaces: Explorations into Mediated Obscurity*. Farnham: Ashgate.

Lovink, G. (2011). *Networks without a Cause: A Critique of Social Media*. Cambridge: Polity Press.

Löwgren, J. and Reimer, B. (2013). *Collaborative Media: Production, Consumption and Design Interventions*. Cambridge, MA: MIT Press.

Luckmann, T. and Berger, P.L. (1964). Social mobility and personal identity, *Archives Européennes de Sociologie* 5(2): 331–44.

Lundby, K. (2014a). Mediatization of communication. In Lundby, K. (ed.) *Mediatization of Communication (Handbooks of Communication Science, Vol. 22)*. Berlin: De Gruyter Mouton.

Lundby, K. (ed.) (2014b). *Mediatization of Communication (Handbooks of Communication Science, Vol. 22)*. Berlin: De Gruyter Mouton.

Lunt, P. and Livingstone, S. (2016). Is "mediatization" the new paradigm for our field? A commentary on Deacon and Stanyer (2014, 2015) and Hepp, Hjarvard and Lundby (2015), *Media, Culture & Society* 38(3): 462–70.

Lyon, D. (2007). *Surveillance Studies:* An Overview. Cambridge: Polity Press.

McBride, C. (2013). *Recognition*. Cambridge: Polity Press.

MacCannell, D. (1976). *The Tourist: A New Theory of the Leisure Class*. London: Macmillan.

McNay, L. (2008). *Against Recognition*. Cambridge: Polity Press.

McQuire, S. (2016). *Geomedia: Networked Cities and the Future of Public Space*. Cambridge: Polity Press.

Madianou, M. and Miller, D. (2012). *Migration and New Media: Transnational Families and Polymedia*. London: Routledge.

Marvin, C. (1988). *When Old Technologies Were New: Thinking About Electric Communication in the Late 19th Century*. New York: Oxford University Press.

Marwick, A.E. (2012). The public domain: Social surveillance in everyday life, *Surveillance & Society* 9(4): 378–93.

Marwick, A.E. (2013). *Status Update: Celebrity, Publicity, and Branding in the Social Media Age*. New Haven, CT: Yale University Press.

Massey, D. (1991). A global sense of place, *Marxism Today*, June: 24–29.

Mau, S., Mewes, J. and Zimmerman, A. (2008). Cosmopolitan attitudes through transnational practices?, *Global Networks* 8(1): 1–24.

Mazzoleni, G. and Schulz, W. (1999). Mediatization of politics: A challenge for politics?, *Political Communication* 16: 247–61.

Mejias, U.A. (2013). *Off the Network: Disrupting the Digital World*. Minneapolis, MN: University of Minnesota Press.

Meuleman, R. and Savage, M. (2013). A field analysis of cosmopolitan taste: Lessons from the Netherlands, *Cultural Sociology* 7(2): 230–56.

Miller, D. (2008). *The Comfort of Things.* Cambridge: Polity Press.

Millington, N. (2013). Post industrial imaginaries: nature, representation, and ruin in Detroit, Michigan, *International Journal of Urban and Regional Research* 37: 279–96.

Milner, A. (1994). Cultural materialism, culturalism and post-culturalism: The legacy of Raymond Williams, *Theory, Culture & Society* 11(1): 43–73.

Molz, J.G. (2007). Cosmopolitans on the couch: Mobile hospitality and the Internet. In Gibson, S. and Molz, J.G. (eds.) *Mobilizing Hospitality: The Ethics of Social Relations in a Mobile World.* Aldershot: Ashgate.

Molz, J.G. (2012). *Travel Connections: Tourism, Technology and Togetherness in a Mobile World.* London: Routledge.

Molz, J.G. (2014). Collaborative surveillance and technologies of trust: Online reputation systems in the "new" sharing economy. In Jansson, A. and Christensen, M. (eds.) *Media, Surveillance and Identity: Social Perspectives.* New York: Peter Lang.

Moores, S. (1988). "The box on the dresser": Memories of early radio and everyday life, *Media, Culture and Society* 10(1): 23–40.

Moores, S. (1993). *Interpreting Audiences.* London: Sage.

Moores, S. (2012). *Media, Place and Mobility.* Basingstoke: Palgrave Macmillan.

Moores, S. and Metykova, M. (2010). "I didn't realize how attached I am": On the environmental experiences of trans-European migrants, *European Journal of Cultural Studies* 13(2): 171–89.

Morley, D. (1992). *Television, Audiences and Cultural Studies.* London: Routledge.

Morley, D. (2000). *Home Territories: Media, Mobility and Identity.* London: Routledge.

Morley D. (2009). For a materialist, non-media-centric media studies, *Television & New Media* 10(1): 114–16.

Morris, J.W. (2015). Curation by code: infomediaries and the data mining of taste, *European Journal of Cultural Studies* 18(4–5): 446–63.

Mosco, V. (2004). *The Digital Sublime: Myth, Power and Cyberspace.* Cambridge, MA: MIT Press.

Mott, C. and Roberts, S.M. (2014). Not everyone has (the) balls: urban exploration and the persistence of masculinist geography, *Antipode* 46(1): 229–45.

Munar, A.M. and Jacobsen, J.K.S. (2014). Motivations for sharing tourism experiences through social media, *Tourism Management* 43: 46–54.

Munt, I. (1994). The "other" postmodern tourism: culture, travel and the new middle classes, *Theory, Culture & Society* 11(3): 101–23.

Myles, J.F. (2004). From doxa to experience: Issues in Bourdieu's adoption of Husserlian phenomenology, *Theory, Culture and Society* 21(2): 91–107.

Nava, M. (2007). *Visceral Cosmopolitanism: Gender, Culture and the Normalisation of Difference.* Oxford: Berg.

Ninjalicious (2005). *Access All Areas: A User's Guide to the Art of Urban Exploration.* Toronto: Infilpress.

O'Neill, S. and Smith, N.H. (eds.) (2012). *Recognition Theory as Social Research: Investigating the Dynamics of Social Conflict.* Basingstoke: Palgrave Macmillan.

Osman, S. (2011). *The Invention of Brownstone Brooklyn: Gentrification and the Search for Authenticity in Postwar New York.* Oxford: Oxford University Press.

Paasonen, S. (2015). As networks fail: affect, technology and the notion of the user, *Television and New Media* 6(8): 701–16.

Papacharissi, Z. (2009). The virtual geographies of social networks: A comparative analysis of Facebook, LinkedIn and A Small World, *New Media & Society* 11(1–2): 199–220.

Papacharissi, Z. (2015). *Affective Publics: Sentiment, Technology and Politics*. Oxford: Oxford University Press.

Papastergiadis, N. (2012). *Cosmopolitanism and Culture*. Cambridge: Polity Press.

Parikka, J. (2012). *What Is Media Archeology?* Cambridge: Polity Press.

Pariser, E. (2011). *The Filter Bubble: What the Internet Is Hiding from You*. New York: Penguin.

Parks, L. (2007). Points of departure: The culture of US airport screening, *Journal of Visual Culture* 6(2): 183–200.

Pattaroni, L. and Adly, H. (2013). Boundaries and urban worlds: The contested ethnoscape of expatriates in Geneva. Paper presented at the International Sociological Association (RC 21) Conference, Berlin, 29–31 August.

Peterson, R.A. (1992). Understanding audience segmentation, from elite and mass to omnivore and univore, *Poetics* 21(4): 243–58.

Phillips, M. (2002). The production, symbolisation and socialisation of gentrification: Impressions from two Berkshire villages, *Transactions of the Institute of British Geographers* 27(3): 282–308.

Phillips, M. (2004). Other geographies of gentrification, *Progress in Human Geography* 28(1): 5–30.

Phillips, M. (2005). Differential productions of rural gentrification, *Geoforum* 36: 477–94.

Phillips, M. (2010). Counterurbanisation and rural gentrification: An exploration of the terms, *Population, Space and Place* 16(6): 539–58.

Pichler, F. (2008). How real is cosmopolitanism in Europe?, *Sociology* 42(6): 1107–26.

Pinder, D. (2013). Dis-locative arts: mobile media and the politics of global positioning, *Continuum: Journal of Media and Cultural Studies* 27(4): 523–41.

Polson, E. (2016). *Privileged Mobilities: Professional Migration, Geo-Social Media and a New Global Middle Class*. New York: Peter Lang.

Portwood-Stacer, L. (2013). Media refusal and conspicuous non-consumption: The performative and political dimensions of Facebook abstention, *New Media & Society* 15(7): 1041–57.

Poster, M. (1995). *The Second Media Age*. Cambridge: Polity.

Rawolle, S. and Lingard, B. (2010). The mediatization of the knowledge based economy: An Australian field based account. *Communications: The European Journal of Communication Research* 35(3): 269–86.

Rawolle, S. and Lingard, B. (2014). Mediatization and education: A sociological account. In Lundby, K. (ed.) *Mediatization of Communication (Handbooks of Communication Science, Vol. 21)*. Berlin: De Gruyter Mouton.

Rérat, P. and Lees, L. (2010). Spatial capital, gentrification and mobility: Evidence from Swiss core cities, *Transactions of the Institute of British Geographers* 36: 126–42.

Riesman, D. (1950/2001). *The Lonely Crowd*. New Haven, CT: Yale University Press.

Ritzer, G. (1999). *Enchanting a Disenchanted World: Revolutionizing the Means of Consumption*. Thousand Oaks, CA: Pine Forge Press.

Ritzer, G. (2015). Automating prosumption: The decline of the prosumer and the rise of the prosuming machines, *Journal of Consumer Culture* 15(3): 407–424.

Robinson, P. (2015). Conceptualizing urban exploration as beyond tourism and as anti-tourism, *Advances in Hospitality and Tourism Research* 3(2): 141–64.

Rofe, M. (2003). "I want to be global": Theorising the gentrifying class as an emergent elite global community, *Urban Studies* 40(12): 2511–26.

Rose, G. (forthcoming). Look Inside™: Visualizing the smart city. In Fast, K., Jansson, A., Lindell, J., Ryan Bengtsson, L. and Tesfahuney, M. (eds.) *Geomedia Studies: Spaces and Mobilities in Mediatized Worlds*. London: Routledge.

Salazar, N. (2010). Tourism and cosmopolitanism: A view from below, *International Journal of Tourism Anthropology* 1(1): 55–69.

Salazar, N. (2011). The power of imagination in transnational mobilities, *Identities* 18(6): 576–98.

Sandercock, L. (2000). When strangers become neighbours: Managing cities of difference, *Planning Theory & Practice* 1(1): 13–30.

Sandercock, L. (2003). *Cosmopolis II: Mongrel Cities in the 21st Century*. London: Continuum.

Sandercock, L. (2010). *Multimedia Explorations in Urban Policy and Planning: An Exploration of the Next Frontier*. Dordrecht: Springer.

Savage, M. (2010). The politics of elective belonging, *Housing, Theory and Society* 27(2): 115–35.

Savage, M. and Gayo-Cal, M. (2011). Unravelling the omnivore: A field analysis of contemporary musical taste in the United Kingdom, *Poetics* 39(5): 337–57.

Savage, M., Bagnall, G. and Longhurst, B. (2005). *Globalization and Belonging*. London: Sage.

Savage, M., Barlow, J., Dickens, P. and Fielding, T. (1992). Culture, consumption and lifestyle. In Savage, M., Barlow, J., Dickens, P. and Fielding, T. (eds.) *Property, Bureaucracy and Culture: Middle-Class Formation in Contemporary Britain*. London: Routledge.

Scannell, P. (1996). *Radio, Television and Modern Life*. Oxford: Wiley Blackwell.

Schivelbusch, W. (1987). *The Railway Journey: The Industrialization and Perception of Time and Space*. Berkeley, CA: University of California Press.

Schulz, W. (2004). Reconstructing mediatization as an analytical concept, *European Journal of Communication* 19(1): 87–101.

Schulze, G. (1995). *Die Erlebnisgesellschaft: Kultursoziologie der Gegenwart*. Frankfurt: Campus.

Schütz, A. (1962). *Collected Papers, Vol 1: The Problem of Social Reality*. The Hague: Martinus Nijhoff.

Schütz, A. and Luckmann, T. (1973). *The Structures of the Life-World*. Evanston, IL: Northwestern University Press.

Seamon, D. (1979). *A Geography of the Lifeworld: Movement, Rest, Encounter*. London: Croom Helm.

Self, W. (2014). Give the freedom of the city to our urban explorers, *Evening Standard*, 25 April. Available at: http://www.standard.co.uk/comment/will-self-give-the-freedom-of-the-city-to-our-urban-explorers-9286780.html.

Shannon, C.E. and Weaver, W. (1949/1963). *A Mathematical Theory of Communication*. Urbana, IL: University of Illinois Press.

Sheller, M. and Urry, J. (2006). The new mobilities paradigm. *Environment and Planning* 38: 207–26.

Silverstone, R. (1994). *Television and Everyday Life*. London: Routledge.

Silverstone, R. (2007). *Media and Morality: On the Rise of Mediapolis*. Cambridge: Polity Press.

Silverstone, R., Hirsch, E. and Morley, D. (1992). Information and communication technologies and the moral economy of the household. In Silverstone, R. and Hirsch, E. (eds.) *Consuming Technologies: Media and Information in Domestic Spaces*. London: Routledge.

Simmel, G. (1900/1990). *The Philosophy of Money*. London: Routledge.

Simmel, G. (1911/1997). The Adventure. In Frisby, D. and Featherstone, M. (eds.) *Simmel on Culture*. London: Sage.

Slager, E.J. (2013). Touring Detroit: Ruins, Representation and Redevelopment. MA thesis in Geography, University of Oregon.

Slater, T. (2006). The eviction of critical perspectives from gentrification research, *International Journal of Urban and Regional Research* 30(4): 737–57.

Smith, N. (2002). New globalism, new urbanism: Gentrification as global urban strategy, *Antipode* 34(3): 427–50.

Soja, E. (2010). *Seeking Spatial Justice*. Minneapolis: University of Minnesota Press.

Soukup, C. (2012). The postmodern ethnographic flaneur and the study of hyper-mediated everyday life, *Journal of Contemporary Ethnography* 42(2): 226–54.

Spigel, L. (1992). *Make Room For TV: Television and the Family Ideal in Postwar America*. Chicago: University of Chicago Press.

Steiner, C.J. and Reisinger, Y. (2005). Understanding existential authenticity, *Annals of Tourism Research* 33(2): 299–318.

Stjernborg, V., Tesfahuney, M. and Wretstrand, A. (2015). The politics of fear, mobility, and media discourses: A case study of Malmö, *Transfers* 5(1): 7–27.

Striphas, T. (2015). Algorithmic culture, *European Journal of Cultural Studies* 18(4–5), 395–412.

Szerszynski, B. and Urry, J. (2002). Cultures of cosmopolitanism, *Sociological Review* 50: 461–81.

Tacchi, J. (1998). Radio texture: Between self and others. In Miller, D. (ed.) *Material Cultures: Why Some Things Matter.* Chicago: Chicago University Press.

Taylor, C. (1989). *Sources of the Self: The Making of Modern Identities*. Cambridge: Cambridge University Press.

Tegtmeyer, L. (2016). Tourism aesthetics in *ruinscapes*: Bargaining cultural and monetary values of Detroit's negative image, *Tourist Studies* 16(4): 462–77.

Thrift, N. (1996). *Spatial Formations*. London: Sage.

Thurlow, C. and Jaworski, A. (2006). The alchemy of the upwardly mobile: Symbolic capital and the stylization of elites in frequent-flyer programmes, *Discourse & Society* 17(1): 99–135.

Tomlinson, J. (1999). *Globalization and Culture*. Cambridge: Polity Press.

Tomlinson, J. (2007). *The Culture of Speed: The Coming of Immediacy*. London: Sage.

Tuan, Y.-F. (1974). *Topophilia: A Study of Environmental Perception, Attitudes, and Values.* New York: Columbia University Press.

Tuan, Y.-F. (1977). *Space and Place: The Perspective of Experience*. Minneapolis, MN: University of Minnesota Press.

Turkle, S. (2011). *Alone Together: Why We Expect More from Technology and Less from Each Other*. New York: Basic Books.

Urry, J. (1988). Cultural change and contemporary holiday-making, *Theory, Culture & Society* 5(1): 35–55.

Urry, J. (1990). *The Tourist Gaze*. London: Routledge.

Urry, J. (2007). *Mobilities*. Cambridge: Polity Press.

Van Dijck, J. (2012). Facebook as a tool for producing sociality and connectivity, *Television and New Media* 13(2): 160–76.

Van Dijck, J. (2013). *The Culture of Connectivity: A Critical History of Social Media*. Oxford: Oxford University Press.

Van Dijck, J. and Poell, T. (2013). Understanding social media logic, *Media and Communication*, 1(1): 2–14.

Van Eijck, K. (1999). Socialization, education, and lifestyle: How social mobility increases the cultural heterogeneity of status groups, *Poetics* 26(5–6): 309–28.

Virilio, P. (1995). *The Art of the Motor*. Minneapolis, MN: University of Minnesota Press.

Walzer, M. (1987). *Interpretation and Social Criticism*. Cambridge, MA: Harvard University Press.

Wang, N. (1999). Rethinking authenticity in tourism experience. *Annals of Tourism Research* 26(2): 349–70.

Werbner, P. (1999). Global pathways: Working-class *cosmopolitans* and the creation of transnational ethnic worlds, *Social Anthropology* 7(1): 17–35.

Wiley, S.B.C., Moreno, T. and Sutko, D. (2012). Assemblages, networks, subjects: A materialist approach to the production of social space. In Packer, J. and Wiley, S.B.C. (eds.) *Communication Matters: Materialist Approaches to Media, Mobility and Networks*. New York: Routledge.

Wilken, R. (2010). A community of strangers? Mobile media, art, tactility and urban encounters with the Other, *Mobilities* 5(4): 449–68.

Wilken, R. and Goggin, G. (eds.) (2015). *Locative Media*. London: Routledge.

Williams, R. (1961/1965). *The Long Revolution*. Harmondsworth: Penguin.

Williams, R. (1974). *Television: Technology and Cultural Form*. London: Fontana.

Williams, R. (1977). *Marxism and Literature*. Oxford: Oxford University Press.

Williams, R. (1980). *Problems in Materialism and Culture: Selected Essays*. London: New Left Books.

Williams, R. (1989). *Resources of Hope: Culture, Democracy, Socialism*. London: Verso.

Wilson, H.J. (2012). You, by the numbers: Better performance through self-quantification, *Harvard Business Review*, September.

Wise, J.M. (2012). Attention and assemblage in the clickable world. In Packer, J. and Wiley, S.B.C. (eds.) *Communication Matters: Materialist Approaches to Media, Mobility and Networks*. London: Routledge.

Wittel, A. (2001). Toward a network sociality, *Theory, Culture & Society* 18(6): 51–76.

Wyly, E. and Hammel, D. (1999). Islands of decay in seas of renewal: Housing policy and the resurgence of gentrification, *Housing Policy Debate* 10(4): 711–71.

Wynne, D. and O'Connor, J. (1998). Consumption and the postmodern city, *Urban Studies* 35: 865–88.

Zukin, S. (1982). *Loft Living: Culture and Capital in Urban Change*. Baltimore, MD: The Johns Hopkins University Press.

Zukin, S. (2008). Consuming authenticity, from outposts of difference to means of exclusion, *Cultural Studies* 22(5): 724–48.

Zukin, S., Trujillo, V., Frase, P., Jackson, D., Recuber, T. and Walker, A. (2009). New retail capital and neighborhood change: Boutiques and gentrification in New York City, *City & Community* 8(1): 47–64.

INDEX

Note: page numbers in bold refer to tables and page numbers in italics refer to figures.